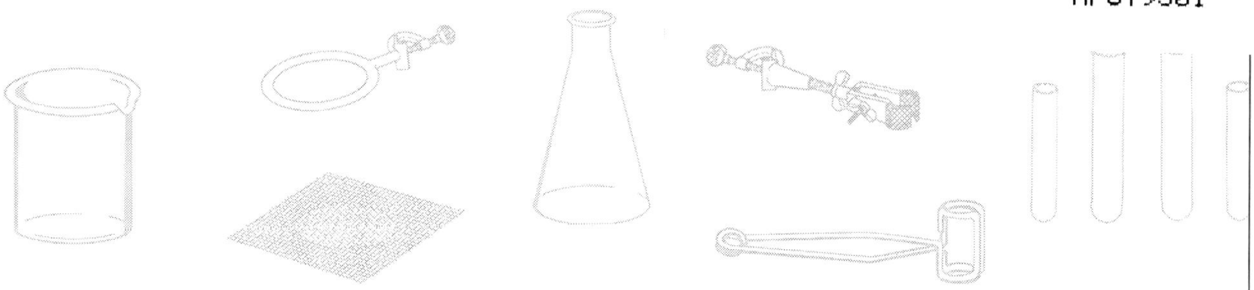

CHEMISTRY the Waldorf Way

Grades 7–8

A Manual for Teachers

Roberto Trostli

A resource companion to ***Physics the Waldorf Way***

LYSTRA BOOKS
& Literary Services

Chemistry the Waldorf Way, Grades 7 & 8: A Manual for Teachers

Contact him at rtrostli@gmail.com.

ISBN: 978-1-7363055-4-6
Library of Congress Control Number: 2021909737

Excerpts from *Elementary Chemistry* and *Natural History* by Eugen Kolisko are reproduced by kind permission of Kolisko Archive Publications, www.koliskoarchive.com

Book design by Kelly Prelipp Lojk

LYSTRA BOOKS & Literary Services

Editorial work and publication management by
Lystra Books & Literary Services, LLC
391 Lystra Estates Drive, Chapel Hill NC, 27517
lystrabooks@gmail.com.

Printed in the USA

This book is dedicated to my daughter, Rachel, who accompanied me on many of my chemistry adventures, and to my granddaughter Jade Amelia Gillert, who seems to have inherited her mother's indomitable spirit.

Chemistry the Waldorf Way

Preface . . . vii
Introduction . . . viii
The Waldorf Method . . . 1
The Waldorf Chemistry Curriculum for Grades 7 & 8 . . . 5
Practical Advice for Teaching Chemistry . . . 11
Working Safely in Chemistry . . . 19

Seventh Grade Chemistry . . . 27

Week 1: Combustion . . . 29
1. Physical & Chemical Changes . . . 33
2. Chemical Changes: Release of Heat . . . 35
3. Chemical Changes: Release of Gas . . . 36
4. Chemical Changes: Change of Color . . . 37
5. Combustion of Plant, Animal, & Mineral Substances . . . 38
6. The Botany of Flame . . . 39
7. Enhancing Combustion: Using Refined Fuel . . . 41
8. Enhancing Combustion: Adding Oxygen . . . 42
9. Enhancing Combustion: Increasing the Fuel's Exposure to Heat . . . 43
10. Controlled Combustion: The Candle . . . 45

Week 2: Gases Combustion . . . 47
1. Products of Combustion . . . 52
2. Carbon Dioxide Does Not Support Combustion . . . 53
3. Testing for Carbon Dioxide with Limewater . . . 54
4. Testing Breath for Carbon Dioxide . . . 56
5. Properties of Carbon Dioxide: Weight . . . 57
6. Demonstrations with Dry Ice . . . 58
7. Generating Oxygen . . . 60
8. Capturing Oxygen . . . 61
9. Combustion in Oxygen . . . 63
10. Synthesizing Carbon Dioxide . . . 66
11. Decomposing Carbon Dioxide . . . 67

Week 3: Water & the Lime Cycle . . . 68
1. Water Cycle . . . 74
2. Solid, Liquid, Gas . . . 75
3. Determining the Boiling & Freezing Points of Water . . . 76
4. Raising & Lowering the Boiling & Freezing Points of Water . . . 78

5. Testing the Electrical Conductivity of Water and Solutions. 80
6. The Electrical Decomposition of Water 82
7. Limestone: Its Sources and Uses 83
8. Burning Limestone 83
9. Slaking QuickLime 84
10. Making Mortar 85
11. Revealing the Carbon Dioxide Contained in Limestone 86

Week 4: Salts, Acids, & Bases 87
1. Salts 92
2. Physical Properties of Salt 92
3. Dissolution & Precipitation of Salt. 93
4. Properties of Acids & Bases 95
5. The Effects of Acids & Bases on Organic Substances 97
6. Identifying Acids & Bases with an Indicator Solution 99
7. Detecting Electricity Through Taste 101
8. Detecting Electricity with a Galvanometer 102
9. The Voltaic Cell 102
10. Colored Flames of Salts 105

Eighth Grade Chemistry 107

Week 1: Sugar 110
1. History of Sugar 114
2. Types of Sugar 115
3. Sugar & Water 116
4. The Solubility of Sugar in Hot Water 117
5. Heating Sugar 119
6. Controlled Heating of Sugar 120
7. Intense Heating of Sugar. 122
8. Heating Sugar Mixed with Potassium Chlorate 123
9. The Dehydration of Sugar by Sulfuric Acid 125
10. Making Candy 126
11. Testing for Sugar 1 127
12. Testing for Sugar 2 128

Week 2: Starch. 130
1. The History of Starch 135
2. Extracting Potato Starch 136
3. Extracting Wheat Starch 137
4. Extracting Corn Starch. 138
5. Starch & Water 139

6. Working with a Non-Newtonian Fluid . . . 140
7. Heating Starch . . . 142
8. Heating Starch Mixed with Potassium Chlorate . . . 143
9. The Combustibility of Dispersed Starch . . . 145
10. Making Pudding . . . 146
11. Testing for Starch . . . 148
12. Testing Food Substances for Starch . . . 149
13. Converting Start to Sugar Using Salivary Enzymes . . . 150
14. Converting Starch to Sugar Using Hydrochloric Acid . . . 151

Week 3: Fats & Oils . . . 154
1. The History & Use of Fats & Oils . . . 158
2. Sources, Properties, & Extraction of Fats & Oils . . . 158
3. Heating Paraffin. . . . 160
4. Heating Oil 1 . . . 161
5. Heating Oil 2 . . . 162
6. Heating Oil 3 . . . 163
7. The Wickless Candle . . . 165
8. Oil & Water . . . 166
9. Oil, Water, & Detergent . . . 167
10. Emulsions. . . . 168
11. Emulsions in Cooking . . . 168
12. Making Soap . . . 170
13. Making an Oil Lamp . . . 171

Week 4: Protein . . . 173
1. The History & Use of Proteins . . . 176
2. Examining Proteins . . . 176
3. Albumen & Air . . . 177
4. Proteins & Water . . . 178
5. Heating Proteins. . . . 179
6. Making Meringues or Egg Drop Soup. . . . 180
7. Interactions of Proteins with Acids & Bases . . . 182
8. Making Ricotta Cheese . . . 184
9. Testing for Protein . . . 185
10. Testing Unknown Substances to Determine Their Identity . . . 187

Appendix 1: Chemicals & Equipment . . . 189
Appendix 2: Common Laboratory Procedures . . . 191
Appendix 3: Chemical Storage . . . 185
Appendix 4: Excerpts from the Work of Eugen Kolisko . . . 197
Bibliography . . . 202

Preface

Over the years, I have received numerous requests to write a chemistry book that would give the kind of guidance I provided in *Physics the Waldorf Way*, a manual for teachers in grades 6, 7, and 8. I resisted these requests because there are plenty of other resources available, and because I doubted whether I had the background or the knowledge to write such a book. Why did I reconsider?

When I look back at my work as a young teacher, I can well recall how the chemistry books by other Waldorf teachers left me feeling intimidated and inadequate. They assumed a level of familiarity with chemistry equipment and procedures that I didn't have and offered choices that I couldn't evaluate. Reading these books convinced me that I didn't know enough to be able to teach chemistry properly.

Unwilling to accept my limitations, I struggled to learn in order to teach, and eventually I found my way. I made plenty of mistakes—thankfully, none of them injured my students or me—and I learned from them. As I became more experienced, I found that the teachers I worked with in summer courses welcomed my advice. At their insistence, I decided to write a book on teaching chemistry after all.

What qualifies me to do so? I may not have a formal background in chemistry, but I do have extensive experience in teaching chemistry. Originally, I didn't know how to work properly and safely with the substances and the procedures, but I do now. Best of all, I have a good sense of what inexperienced teachers need to know and do, and I am able to present and explain my ideas clearly and concisely.

This book is directed towards teachers and parents who are teaching Waldorf chemistry for the first time. It is the product of many decades of refining the curriculum so that it makes conceptual sense, choosing the demonstrations and activities that reveal the phenomena most vividly, and becoming adept with the Waldorf method so that students truly learn.

I hope that this book gives you everything you need to journey confidently into the uncharted territories of the world of chemistry. And I hope that by working with this book you will develop an ever deeper appreciation for the mystery and meaning of the world around us.

Roberto Trostli
Richmond, VA
Summer 2020

Introduction

This book will give you all the help you need to prepare for your chemistry blocks and will give you the guidance you need to teach chemistry confidently, safely, and effectively. It is not a chemistry textbook but a manual for teaching chemistry in grades 7 and 8.

Here is what it contains:

- *Chapter 1: The Waldorf Method* – the essential principles that make Waldorf Waldorf
- *Chapter 2: The Waldorf Chemistry Curriculum* – an overview with commentary
- *Chapter 3: Practical Advice for Chemistry Teaching* – ideas for working with the parts of the lesson and for the academic and artistic work the students should do
- *Chapter 4: Working Safely in Chemistry* – everything you need to do to insure that you and your students work safely
- *Chapter 5: Seventh Grade* – introduction and background to establish a foundation for the block
- *Chapters 6–9:* demonstrations and lab activities in combustion; gases; water; and salts, acids, and bases
- *Chapter 10: Eighth Grade* – introduction and background to establish the foundation for the block
- *Chapters 11–14:* demonstrations and lab activities in sugar, starch, fats, and proteins
- *Appendices:* essential information regarding equipment, materials, chemical storage, common laboratory procedures, and additional resources

I encourage you to read Chapters 1 through 4 as you begin your preparation because they provide a philosophical and methodological orientation to the work ahead.

I recommend using Chapters 5 through 14 as you prepare for your block. They will provide everything you need.

Before and during your teaching, read and review Chapter 4 to make sure that you are following the safety guidelines.

In order to make this manual as easy to use as possible, each entry includes:

Introductory comments
Equipment
Materials
Procedures
Safety considerations
Comments
Related topics for additional exploration

Space has been left in the margin for you to write notes and reminders.

If anything in this book needs to be clarified or corrected, please contact me at rtrostli@gmail.com

Blessings on your work!

– *Roberto Trostli*

Chapter 1: The Waldorf Method

Waldorf education strives to apply an anthroposophical understanding of the human being to the task of education. That understanding forms the basis for an age-appropriate curriculum and methodology that meets the students' intellectual, emotional, and physical needs. The Waldorf method works best when it is based on the teacher's work with anthroposophy, but it also can stand on its own as an effective way to teach science to middle school students.

In *Physics the Waldorf Way*, I presented the anthroposophical basis for the Waldorf method, and I explained how to apply that method. Anyone planning to teach chemistry will benefit from referring to that text.

What are the essentials of the Waldorf method?

1. **Working with the whole human being**
   According to Rudolf Steiner, human beings incarnate on earth to do what was not possible for them to do in the spiritual world. While on earth, we are beings of body, soul, and spirit and go through life stages in which we develop different aspects of our selves. When we apply this view of the human being to our work with students, we help them fulfill the reason they came down to earth: to become themselves and to prepare for their life's work.

2. **Including activities that address the students' thinking, feeling, and willing**
   In every lesson, we should appeal to our students' thinking, feeling, and willing. The various activities in the lessons develop these capacities in different degrees: the review of the previous day's work calls on the students' thinking; the presentation or experience of the new material appeals to their feeling; and the students' academic and artistic work requires them to apply their will. If we always keep these elements in mind, our lessons will be nourishing as well as instructive, meaningful as well as interesting.

Rudolf Steiner's View of the Human Being

Rudolf Steiner articulated a view of the human being that has both sensible and supersensible "members." These include the **ego**—that imperishable part of our true self; the ***astral body***—our "feeling body," which allows us to have sensations, perceptions, and emotions; the ***etheric body***—our "life body," which is responsible for all of our life functions and also for memory; and our ***physical body***—the mineral basis for our physical existence. For those who are interested in learning more, see Rudolf Steiner's extensive descriptions in ***Theosophy*** or in ***An Outline of Esoteric Science***.

3. **Using the progression of conclusion/judgment/concept to develop the students' thinking**
   The three parts of the science lesson—demonstration/activity, recapitulation, and review—address the different aspects of the students. During the demonstration or activity, students form conclusions, which are a function of the ego. When students hear the recapitulation, they use their imagination and memory to form judgments, which are a result of the work of the astral body. When we review the lesson the following day, students apply their memory and their thinking, which are activities of the etheric body. By using this methodology, we help students fully experience phenomena and strengthen the foundations for clear thinking.

4. **Allowing phenomena to reveal their characteristics and trying to view them as "parables of the spirit"**
   The natural world reveals itself through phenomena that can be explained intellectually, but these phenomena should also speak to our hearts. Young adolescents need to learn what the physical world is trying to teach them, rather than focusing only on explanations and concepts. When we help students to form a living relationship to the world, we strengthen their love for life.

   Rudolf Steiner encouraged teachers to view natural phenomena as "parables of the spirit" that can be discovered through contemplation. For example, we can view combustion as a picture of what happens to us at the portal of death. When something combusts, it yields ash, gases, heat, and light. Viewed as a parable, we can view the ash as what remains of the physical body, the gases as an expression of the soul, and the heat and light as embodiments of the spirit. We are beings of body, soul, and spirit; at death, these three aspects of our being separate and return to their proper homes.

Such parables should not be shared with the students, but our efforts to seek such revelations of the spirit in the physical world help our students view the world less materialistically.

5. **Learning about everything in the context of the human being, the natural world, and the created world**
   In order to help our students become self-aware world citizens who are able to work with others, we need to strengthen their sense of connection to everything in the world. We accomplish this by presenting everything in context and by helping students recognize and appreciate both the visible and the invisible warp and weft of the world. Waldorf education should develop both consciousness and conscience in our students so that they will work towards a more human future.

6. **Committing to relationships**
   Waldorf education depends on human relationships. Children can only learn what it is to be human through their relationships with other human beings who are also striving to realize their full humanity. Young children learn primarily through imitation, grade-school children through loving discipleship, and older students through their respect for adults. At every stage of development, relationships serve as the foundation for the children's journey of becoming. Teachers who commit to their relationships with the students, their parents, and their colleagues provide their students with a worthy model of the developing human being.

7. **Meditating**
   Teachers who engage in a regular practice of anthroposophical meditation will find unexpected sources of inspiration and guidance for their lives and for their work. Rudolf Steiner characterized anthroposophy as "a path of knowledge, to guide the spiritual in the human being to the spiritual in the universe." By following this

path, we are led towards our higher selves, we connect more deeply with the world, and we create a channel for spirit to manifest itself through our thoughts, words, and deeds.

To educate youth
Means to foster and tend
In matter, the spirit;
In today, the tomorrow;
In earthly life,
The Spirit's existence.

– Rudolf Steiner

Chapter 2: The Waldorf Chemistry Curriculum for Grades 7 & 8

As Waldorf teachers, we have the opportunity to help our students develop their individuality, learn to collaborate with others, and develop the capacities to become stewards of the earth. The science lessons in the upper grades assist us in this process. As students learn about themselves and as they connect what they learn in the classroom with the natural and created worlds, they will deepen their relationship to life and strengthen their commitment to make a positive contribution in the future.

Rudolf Steiner gave brief indications for teaching chemistry in his lectures on curriculum included in *Discussions with Teachers*. In these indications, he named the subjects to be covered and stressed the importance of helping students gain a picture of commercial and industrial processes.

Here is what Rudolf Steiner said:

> *[In the seventh grade,] you start from an everyday process such as combustion and try to make the transition to simple concepts of chemistry.*
>
> *We also attempt to apply the concepts the children have acquired in the fields of physics and chemistry to develop a comprehensive view of some specific commercial or industrial processes.*
>
> *[In the eighth grade,] you continue to develop simple concepts of chemistry so that the children also learn to grasp how industrial processes are related to chemical ones. In connection with chemical concepts, you also attempt to develop what needs to be said about the substances that build up organic bodies—starch, sugar, protein, and fat.*
>
> *Once again, you present a comprehensive picture of industrial and commercial relationships as they relate to physics, chemistry, and geography.*

The basic elements of the Waldorf chemistry curriculum in the elementary school were established by Eugen Kolisko, who taught chemistry in the first Waldorf School and worked with Rudolf Steiner. Since then, many other Waldorf teachers have expanded and refined the elementary and high school chemistry curricula.

Unfortunately, all of these curricula require a level of knowledge and experience that many teachers lack. This expectation inspired me to develop a curriculum that would allow less-experienced Waldorf teachers or homeschool parents to teach chemistry with confidence. Those who have more background or experience in the subject will still find my approach useful, but they should modify it according to their students' interests and needs.

The Greek Elements

In my opinion, the ancient Greek elements lie at the heart of the Waldorf curriculum. While modern chemistry identifies 118 chemical elements and innumerable compounds, the ancient Greeks thought that everything in the material world could be expressed by four elements: earth, water, air, and fire. They did not view these elements only as the substances for which they were named but also as archetypes of qualities and processes.

The Greek elements correspond to the states of matter: earth to solid, water to liquid, and gas to air. Energy, which is not material, corresponds to fire. The seventh-grade curriculum explores these elements.

- **Fire:** When we examine combustion, we see the archetype of fire being expressed. Combustion breaks substances up into their component parts. Energy, which lies hidden in the substance, is unleashed and revealed and expresses itself as heat and light.
- **Air:** The gases of combustion—oxygen and carbon dioxide—express the archetype of air. Gases are typically invisible, fill an enclosed space, and are always in motion. Gases permeate substances and determine their characteristics. These are all properties of the Greek element air.
- **Water:** Water expresses the archetype of the Greek element for which it is named. Its fluidity and mobility are expressed by the water cycle and by the fact that water is almost always in motion. Water's ability

to bring substances into relationship with each other also expresses its archetypal mediating function. Finally, water is considered a universal solvent because it is able to dissolve more substances than any other liquid.

- **Earth:** Salts and their derivatives—acids and bases—express the archetype of earth. Salts are heavy, and some of them are insoluble. Those that are soluble can precipitate out of the solution and resume their solid form. Many salts are unreactive and require very high temperatures to liquefy or vaporize. Acids and bases show two different sides of salt's nature: acids are light and volatile, while bases are heavy and stable. Both acids and bases have a relationship to water, which often binds them into a salt, but the reaction between them can also be accompanied by a release of heat.

The eighth-grade curriculum focuses on food substances. These too are connected to the ancient Greek elements.

- **Water:** Sugar is deeply connected to the ancient element of water. This is shown primarily by its solubility and by the large numbers of substances it can create.
- **Earth:** Starch and cellulose are connected to the ancient element of earth. This is shown by their stability and relative heaviness, and by how they maintain their nature unless they are forced to undergo transformation.
- **Fire:** Fats are connected to the ancient element of fire. This is shown by their volatility and flammability and by the energy they release. When fats burn, they leave almost no ash, showing just how combustible they are.
- **Air:** Proteins are connected to the ancient element of air. Plants manufacture proteins from glucose, but they also need nitrogen for this process. Plants absorb that nitrogen from the soil. They cannot absorb it directly from the atmosphere, even though the atmosphere is primarily composed of nitrogen.

Photosynthesis—the process by which plants utilize sunlight to transform water and carbon dioxide into sugar and oxygen—forms the basis for all life on earth. The plant converts sugar into ever more complex substances—starches, cellulose, fats, and proteins—and we will follow this progression in our chemistry block.

Greek Elements: Grade 7

topic	element
combustion	fire
gases	air
water	water
salts, acids & bases	earth

Greek Elements: Grade 8

topic	element
sugar	water
starch & cellulose	earth
fats & oils	fire
proteins	air

Understanding the World Around Us

In addition to listing suitable topics for the seventh and eighth grades, Rudolf Steiner also indicated that teachers should show how concepts in the physical sciences are applied in commerce and industry. In *A Modern Art of Education*, he explained why this is so important:

> *Consider how many people there are today who get into a train without the faintest idea of how it operates or what makes it move. They see a train every day and have absolutely no idea of the mechanics of a locomotive. This means that people are surrounded by inventions of the human mind but have no contact at all with them. It is the beginning of an antisocial life simply to accept inventions of the human mind without at least understanding them in a general way. In a Waldorf school, therefore, when the children reach fourteen or fifteen, we provide instruction and experience in matters that play a role in practical life.*

Today we are surrounded by innumerable inventions and processes that we take for granted but do not understand. As Waldorf teachers, we have a duty to help students find a connection to these human creations so that they will feel more at home in the world and will be better able to cope with the challenges the world presents.

Differences between the study of physics and chemistry

Since both physics and chemistry are taught in the upper elementary grades, it behooves us to consider the major differences between them.

- Physics is the study of forces, while chemistry is the study of substances and their interactions.

- Forces reveal themselves through their results, and it's usually possible to determine cause and effect, whereas substances react with one another based on their chemical structure, which is not perceptible.
- Physics phenomena are usually easier to understand than chemical phenomena, which are often more mysterious, more engaging, and more beautiful.
- Phenomena in physics can often be paused in order to investigate what is happening; chemical processes and reactions tend to run their course and can usually not be interrupted.
- Physical changes and reactions are often reversible, which makes them more predictable; chemical reactions involve changes whose causes may remain hidden.
- Physics tends to be more familiar and less dangerous than chemistry. Students may feel some trepidation about studying chemistry, but they usually anticipate it more eagerly than physics.
- The study of physics is more directed to the thinking and to the will, while the study of chemistry appeals more to the feelings.
- In seventh grade, students tend to be familiar with the physical forces they study and less familiar with the topics they study in chemistry.
- In eighth grade, the opposite is true; students already know a great deal about the chemistry of foodstuffs, while they know less about fluid mechanics and aeromechanics.
- It may be easier to find the connections among chemical phenomena and the human being, the natural world, and commercial or industrial applications than it is in physics phenomena.

Chapter 3: Practical Advice for Teaching Chemistry

Background Reading

This manual should give teachers and homeschool parents who do not have a specialized knowledge of chemistry all the information they need to present demonstrations and laboratory activities that reveal chemical phenomena. Teachers and parents should, however, inform themselves further about the subjects that they are presenting, because they need to be able to extend the students' view beyond the classroom.

Anthroposophical authors provide us with valuable insights into the subjects we teach. To give a sense of what these types of sources offer, I have included excerpts from the works of Eugen Kolisko in Appendix 4. Although you may read anthroposophical works as a way of broadening your perspective, I encourage you to be scrupulous about which views to share with your students and their parents to avoid misunderstandings and distrust.

Demonstrations

- **Materials & Equipment:** Laboratory science depends on materials and equipment, and teachers need to familiarize themselves with the equipment and procedures they will be using. A glossary of laboratory materials and equipment is provided in Appendix 1, and a description of common laboratory procedures is provided in Appendix 2. Study those appendices as you read through this book and prepare to teach through demonstrations and laboratory activities. It is important that teachers use and teach the students the proper vocabulary for substances, equipment, and procedures so that their students are properly prepared for their high school science courses.
- **Practice:** In order to perform demonstrations and lab activities properly and confidently, teachers need to practice. Practice the demonstration at least three times. Though three practices may sound excessive, it is actually the minimal number. Musical performers and actors practice hundreds of times in order to perform

with confidence and to be able to allow the meaning and the message of the piece to be expressed through them.

- **Demonstration table:** Obtain a demonstration table with a waterproof top, or use a table or desk covered with heavy-duty aluminum foil. Alternatively, you can use a food service cart that has two or three levels. The cart has the advantage of allowing you to store some of the items you're going to need in the course of demonstrations.
- **Visibility:** Visibility is of prime importance. Having a demonstration table that is a little taller than a desk is helpful. Students should gather around the table. Some can sit, others stand, and others can sit on desks.

  Do demonstrations using the largest equipment available so that everyone can see the phenomena. For instance, instead of using a test tube to demonstrate how cabbage juice indicates the presence of an acid or a base, use a large flask, graduated cylinder, or clear glass vase.
- **Demonstration technique:** Formally introduce the demonstration, and tell the students in general terms what they should watch for. Only talk during the demonstration when it is absolutely necessary, and challenge the students to keep their reactions and comments to themselves. If the demonstration has more than one part, introduce each part separately.

  When a demonstration or lab activity doesn't go as you expected, it is usually because of:
  - Faulty instructions
  - Faulty materials or equipment
  - Faulty execution

  Teachers can address all three of these causes by proper preparation, which includes:
  - Trying out the instructions to see if they are correct and if not, correcting them
  - Working with the materials and equipment to see if they perform as expected
  - Practicing the procedures to make sure they work and to determine if students need to be alerted to possible challenges

If these three areas have been covered, the teacher can be confident that everything will almost always go as planned. If something does go awry, the teacher needs to take responsibility and show the students how to deal with mistakes by determining the cause and making the necessary corrections. We all make mistakes, but they should not be the result of laziness or carelessness. To keep mistakes to the absolute minimum, we need to do the necessary preparation and take the necessary care.

Laboratory Activities

Lab activities take much more time and preparation than demonstrations, but they are important, especially for eighth graders, who need to work independently whenever possible.

- **Preparation & Practice:** Like demonstrations, lab activities require preparation and practice. If teachers do justice to these processes, labs will go more smoothly, students will find them more satisfying, and they will learn more.
- **Time:** Allow more time for lab activities than you think they may require. Students need time to review the instructions and to ask questions. They need time to gather their materials and set them up. They need time to perform the lab carefully and to note their observations. And don't forget cleanup, which also takes time. It is far better to have a little extra time at the end of the activity than to rush to finish.
- **Groups:** In seventh grade, lab groups of three are ideal. Larger groups tend to engender passivity in some of their members, and sometimes members form their own sub-groups. In eighth grade, students can work in groups of three or in pairs. Sometimes, I have students work with the same partner for a whole week to engender greater collaboration.

  When groups are assigned, students should decide who will gather the materials, who will perform the activity, and who will take notes. If the activity has more than one part, they also need to decide whether and when the roles should change. Everyone in the group should be responsible for cleanup.
- **Materials & Equipment:** In order for a lab to work successfully, each group needs to have the materials and equipment it needs. There should also be a reserve of the substances being used in case students need additional materials to repeat or to do additional procedures.

The materials and equipment should be laid out neatly. This can either take the form of an activity kit that contains everything laid out on a tray, or the materials and equipment can be arranged in one location so that each group can take what it needs. To insure that students have everything they need, provide them with written lab instructions.

- **Instructions:** When all the groups have gathered all their materials and equipment, review the instructions and answer students' questions. Make sure that everyone is clear about what to do before anyone starts. In the eighth grade, I usually give written instructions and leave time to review them and to answer questions. If the lab has several parts, indicate this on the instructions or the board so that the lab can proceed without interruption.
- **Mood & Conduct:** Insist on a quiet and diligent mood. The laboratory is not a place to play, and the equipment should be treated like tools, not toys or weapons. Students can talk about what they're doing, but I encourage them not to voice their observations because everyone should observe for him or herself. I usually wander around the room observing the labs, sometimes directing students' attention to something if they're not paying attention, and I try to resist the urge to say too much.
- **Cleanup:** Cleanup plans need to be figured into lab activity time. It's important for the students to participate in the cleanup. The teacher should not clean up after them.

Recapitulations

The recapitulation contains a brief factual account of what was done and a description of what occurred. It is given by the teacher and should be done as soon as the demonstration or lab activity is over.

- **Method:** The recapitulation can be introduced by saying, "Now let's review in our minds what was done and what occurred." It's important to keep the recapitulation as objective as possible so that the students can form accurate memory pictures. Describe the procedures and phenomena, but don't explain or use conceptual language—that can come during the review. Use the passive voice as much as possible so that the students' minds stay focused on the phenomenon, not on whoever performed the demonstration or activity. It is important to avoid illustrating the procedures through gestures or pointing to the equipment so that the students create purely imaginative pictures of what is being described.

- **Importance:** The recapitulation is a vital part of learning because it begins the process of raising sensory impressions into the imaginative and cognitive realms. The imaginative pictures that students form during the recapitulation are taken into sleep, where they are enhanced by the students' experiences in the spiritual world before they awake. In the morning, the students will have a clearer, more profound recollection of their experiences, which will form the basis for understanding concepts.

Reviews

- **Purpose:** The review allows the students to share their recollections, to reflect on their experiences, and to articulate their understanding of the phenomenon or concept. The first part of the review should account for the topic, the materials, the procedures, the observations, and the results. It should lead to a clear formulation of the characteristics revealed by the demonstration or lab activity. The second part of the review should focus on explanation, interpretation, and meaning. Students should tell about other experiences or prior knowledge of the phenomenon, and they should consider the meaning of what they have learned.
- **Method:** I typically begin the review by posing a series of questions for students to answer in their notebooks. These questions focus on what was done and what occurred. Students should try to account for everything they experienced and observed. It is important for students to have a chance to see whether they are able to remember the basic facts of the phenomenon. If they don't remember something, they can fill it in when we review the answers. For the second part of the review, I pose open-ended questions and topics for discussion. This part of the review can—and should—digress and be far-reaching, generating more questions than answers.

  Re-presenting a phenomenon during the review: Sometimes students didn't notice or don't remember details—or even essential aspects—of a demonstration or lab activity. Rather than simply having them listen while their classmates share their recollections, we may decide to present the phenomenon again. In this case, direct the students' attention to aspects of the phenomenon that were in question.

Main Lesson Schedule

I suggest the following schedule for a two-hour main lesson.

1. Warm-up exercises . . . . 15–20 minutes
2. Review . . . . 20+ minutes
3. Work time . . . . 50+ minutes
4. Demonstration & Recapitulation . 20 minutes

When there is a lab activity, adjust as follows:

1. Warm-up exercises . . . . 15–20 minutes
2. Review . . . . 20+ minutes
3. Work time . . . . 30 minutes
4. Activity & Recapitulation . . . . . 40+ minutes

Materials & Equipment

In order to teach chemistry effectively, you need to have access to all the materials and equipment for demonstrations and for the students' lab activities. These materials should be gathered well in advance, because you may not be able to locate or obtain them at the last minute. One of the reasons to rehearse is to assess whether you have everything you need. Keeping everything neat and orderly teaches the students respect for the equipment they use. Storing and working with the materials safely protects the students, the teacher, and the school.

Some of these materials can be dangerous and must be stored and handled carefully. For specific instructions for chemical storage, see Appendix 2.

Main Lesson Work

No matter how engaging a demonstration or lab activity or how meaningful a review, the academic and artistic work time is the most important part of the lesson. Because chemistry is such a visual subject, and the characteristics and concepts can be captured so succinctly, I recommend that students use a highly graphic approach to their main lesson book work and try to capture the essence of each subject on a single page. Larger paper, such as 11-inch-by-14-inch sheets, gives students greater flexibility for layout and more options for illustrations and text.

When designing their pages, students should consider the following elements in terms of how well they convey the subject:

- page layout
- types of lettering
- illustrations, diagrams, drawings, or cartoons
- media—pen, pencil, colored pencil, crayon, pastel, watercolor
- economical use of text

While the main lesson book has many virtues, eighth graders appreciate being able to use other media and formats. I think that they should be encouraged to use the chemistry block as an opportunity to create:

- a poster or set of posters
- a collage or set of collages
- a cartoon sequence
- a drawing book
- a video presentation
- a narrated slide show
- an invention or practical project
- an illustrated children's book

In addition to their main lesson book or project, students should keep a notebook in which they:

- write reflections or responses to review questions
- take notes during the review or during labs
- compose rough drafts
- sketch ideas for their artistic work or main lesson book page designs

Although this notebook is not as polished a product as the main lesson book, it provides a valuable record of a student's experience of the main lesson block.

The Scientific Method and Lab Reports

In high school and college, students will be expected to be familiar with the scientific method and with lab reports. Therefore, plan to include a brief description of the scientific method on several occasions in seventh and eighth grade and give students a few opportunities to write lab reports. The typical lab report contains the following elements:

- Question/Problem
- Hypothesis
- Materials
- Procedure
- Observations
- Conclusions

In my opinion, there is no need to do more than one lab report per block. Although it is an economical way to communicate scientific information, it does not allow students to express their connection to and appreciation for the subject.

Chapter 4: Working Safely in Chemistry

Some substances used in chemistry experiments are dangerous and some of their interactions hazardous. As a result, teachers need to take extra measures to make sure that they are protecting themselves, their students, and their school. Teachers are encouraged to educate themselves as to the properties of these substances before working with them. While the author of this text has suggested demonstrations and activities that have been tested for safety and for which safety instructions are set out in the text, neither the author nor the publisher is responsible for any damages that may result from these demonstrations and activities.

There are three essential requirements for working safely with chemistry:

1. Know the properties of the substances you are working with
2. Know the methods and procedures that are necessary for working safely with the equipment and materials
3. Know what to do when something goes wrong or if there is an emergency

1. Properties of Substances

Although most of the substances used in the chemistry activities in this manual are common household materials, a few of them do require special handling. The most dangerous substances (corrosives and flammables) should only be handled by the teacher.

Teachers need to thoroughly inform themselves about dangerous substances by reading the Material Safety Data Sheet that is provided when you purchase a chemical. The MSDS provides information about the potential hazards of the chemicals and how to dispose of them safely. Teachers also need to know about the potentially dangerous interactions between chemicals.

Unlike physics, where the equipment and materials are left out for the students to work with on their own, the chemicals and substances used in chemistry must be stored when not in use. This requires the school to provide appropriate storage and for the teachers to maintain it properly.

2. Safe Methods and Procedures

Teachers can insure that everyone remains safe by:

- setting up the laboratory area safely
- requiring personal protective equipment
- insuring adequate ventilation
- maintaining laboratory discipline and good conduct
- using proper laboratory vocabulary and techniques
- mastering demonstrations and activities through practice
- storing chemistry equipment and materials safely
- cleaning up and disposing of chemicals properly

CLASSROOM LABORATORY SETUP

Most Waldorf schools do not have a laboratory where chemistry can be taught; rather, teachers usually have to adapt their classrooms for the chemistry main lesson block. Because the classroom is used for other classes during the rest of the day, one should keep the special adaptations for chemistry to a minimum by:

- using a demonstration table
- having a chemistry cart that contains the day's materials and equipment
- storing the equipment and materials for the week in bins

Demonstration Table: Demonstrations should be performed on a demonstration table in the front or the back of the room, with desks pushed back so that students can stand or sit in a semicircle around the demonstration table. For demonstrations that involve vigorous combustion or reactions, a laboratory safety shield needs to be set up on the demonstration table.

Chemistry Cart: A cart that can hold all the equipment and materials for the day's demonstrations is very useful. Leaving everything on the cart until it is needed allows the demonstration table to be kept clear.

Equipment and Materials Storage: The equipment that will be needed for the week should be gathered in advance and kept in plastic bins in a closet or under a table in the classroom. Using bins reduces the number of trips to the science equipment closet during the course of the week.

Chemicals and materials need to be kept in storage until the morning of the day they are to be used and returned after class so that students don't have access to them.

PERSONAL PROTECTIVE EQUIPMENT

Teachers and students must wear goggles, gloves, and aprons when working with any chemical that could harm them. Although almost all of the chemicals used in the lab activities in this manual are safe, everyone needs to wear PPE for those that could result in skin or eye injuries.

For laboratory work involving dangerous chemicals or reactions, students should wear long-sleeved tops, trousers, and close-toed shoes, and they should tie back their hair.

VENTILATION

In the absence of a lab with a fume hood, teachers need to make arrangements for maintaining good air quality during their chemistry classes and for the rest of the day when the room is used for other classes.

Fortunately, only a few of the demonstrations and activities require special arrangements for maintaining air quality. The best way to address smoke production is to keep the quantities of combustible material small—just large enough to reveal the phenomena. A window fan or a box fan placed in a window should clear the air. Those demonstrations that generate larger amounts of smoke should be done outside. Any demonstration that produces toxic vapors—such as the dehydration of sugar by sulfuric acid—must be done outside with the students standing upwind of the reaction.

LABORATORY DISCIPLINE AND GOOD CONDUCT

Working in a laboratory requires special conduct from students to keep everyone safe. Teachers need to make expectations for students' conduct explicit and to review them with students before the block and then periodically as necessary.

PROPER LABORATORY VOCABULARY AND TECHNIQUES

Many class teachers lack any kind of formal laboratory training, so they will need extra preparation to teach chemistry. This preparation includes learning:

- the proper names for laboratory equipment and materials
- how to use the equipment correctly
- how to perform laboratory procedures properly

Students benefit from learning laboratory vocabulary and from being taught to perform laboratory procedures correctly. It develops their respect for the subject and the activities and gives them confidence that they can learn to work properly. (Appendix 2 has important information about the safe use of common laboratory procedures. Be sure to study these as part of your preparation for demonstrations and lab activities.)

PREPARATION AND PRACTICE

A teacher must commit herself to never doing any chemistry demonstration or activity that she has not practiced at least three, if not more, times. Only through practice can the teacher learn the nature and possible dangers of a chemical procedure or reaction. Practicing develops the teacher's knowledge and confidence. Students will have a much better experience with—and will learn more from—a teacher who knows what to do and how to do it.

STORING CHEMISTRY EQUIPMENT AND MATERIALS SAFELY

A clean, well-organized classroom supports student working and learning. During a chemistry block, the organization of classroom and the storage of equipment and materials must also insure safety.

Much of the equipment and most of the chemicals used in the chemistry block do not require specialized storage, but keeping them organized and storing them safely is a good practice. A few of the chemicals that are used do need specialized storage, and schools must provide the appropriate storage to maintain safety and to protect against litigation. (A list of storage requirements is provided in Appendix 3.)

CLEANUP & DISPOSAL

All of the demonstrations and activities require cleanup, and teachers should create a cleanup routine that works efficiently and safely. Students can be assigned to assist in cleanup after demonstrations on a rotating basis; for lab activities, each group should be responsible for cleaning up its own equipment and materials. Instructions for cleanup should include where and how to put away the equipment and materials.

Almost all of the materials and chemicals used in the demonstrations and activities in this manual can be safely disposed of in the trash or flushed down the sink after they have been diluted and neutralized by the teacher. To dispose of potassium dichromate, Fehling's, Benedict's, and biuret solutions, teachers should consult local ordinances regarding safe disposal.

3. Know What to Do When Something Goes Wrong or When There Is an Emergency

There are five ways to minimize the number of things that might go wrong during chemistry:

- prepare thoroughly
- exercise caution
- plan for the unexpected
- be prepared for injuries
- learn from your mistakes and from students' mistakes

PREPARING THOROUGHLY

Inadequate preparation can have serious repercussions, so teachers need to commit to sufficient preparation for teaching chemistry. A teacher who cannot make or follow through on this commitment should ask for assistance or even consider having someone else teach the subject.

EXERCISING CAUTION

Chemistry is an exciting subject, and that excitement might be amplified by the whiff of danger that surrounds some of the demonstrations and activities. Students will often ask, "Are we going to have any explosions?" I use that as an opportunity to explain to them what an explosion is and to inform them why we will not have any.

I believe that to teach chemistry properly and safely, teachers need to embrace their role as a modest servant of nature, not as a magician or an entertainer. The teacher's role is to reveal or perhaps to surprise, not to shock or amaze.

Teachers can exercise caution during demonstrations by the manner in which they perform a procedure—slowly, carefully, and thoughtfully—and by the amounts of substances they use—as little as necessary.

Students also need to learn to exercise caution in their attitude and their conduct. Teachers should help students understand that this will allow them to learn more and will keep everyone safer. In the chemistry lab, tools may not be used as toys, and nothing should ever be done to surprise, intimidate, or frighten anyone.

Learning to respect caution teaches the students much more than how to work safely in chemistry; it helps them to become conscientious and responsible adults.

PLANNING FOR THE UNEXPECTED

Nature has its own laws and rules, and when we don't know them or heed them, we may be surprised. Preparing properly will minimize the number of unexpected events. When something unexpected happens, though, we need to be able to fall back on a prepared plan of action, rather than figuring out what to do in the moment.

When practicing a demonstration or lab activity, teachers need to think about potential dangers, about what might go wrong, and what to do if something does go wrong. For instance, when burning magnesium, students are going to need to be warned not to look at the bright flame for more than a moment, and the teacher should know why such a warning is needed.

Students need directions that include instructions about how to handle mishaps or accidents. For instance, when eighth graders convert starch to sugar with hydrochloric acid, they need to know that if the acid spills, it has to be neutralized with baking soda. Whether this is done by the teacher or the students depends on the students' level of maturity.

DEALING WITH INJURIES

Although we strive to work safely, sometimes a student will suffer an injury. If proper precautions have been taken, it will most likely be a very minor one, but every injury should be treated.

The teacher needs to have developed an injury response plan and to review it with the students at the beginning of the block and before any demonstration or activity that is potentially dangerous. Having an injury response plan helps prevent injuries by raising everyone's consciousness and strengthening everyone's confidence that if an injury does occur, we know how to respond.

LEARNING FROM MISTAKES

Mistakes teach us what we need to learn. When we heed them, mistakes help us to see what we need to do differently under similar circumstances or how we need to be different to meet our challenges.

All of us make countless little mistakes. If we recognize them and learn from them, we are less likely to repeat them or to require bigger mistakes to teach us what we have not yet learned. Bigger mistakes are life's way of telling us to pay attention to what we need to learn. And the biggest mistakes are life's insistence that we'd better learn this lesson, or else!

Our biggest mistakes have potential ramifications that extend far beyond the classroom. I cannot stress strongly enough: listen and learn from minor mistakes so that you do not place your students and your school in jeopardy.

The following Buddhist proverb will serve you and your students well, both in chemistry and in life: "How you do anything is how you do everything."

Chapter 5: Seventh Grade Chemistry

In seventh grade, students are introduced to two new subjects: physiology and chemistry. In some ways, these subjects are opposite: physiology turns the students' gaze inward to the workings of their bodies; chemistry directs their gaze outward to the substances and processes of the world. In other ways, these subjects reflect and complement each other: the process of digestion shares many similarities with combustion; respiration with the behavior of gases; circulation with the properties of water; and the study of salts provides a foundation for understanding the skeleton and nerve processes. Both physiology and chemistry awaken curiosity and inspire wonder, and they are good antidotes to the typical seventh grader's pose of weary worldliness. Both contain mysteries of existence that can be revealed through careful consideration leavened with contemplation. The seventh-grade teacher is blessed to be the one to introduce students to such amazing new worlds.

Seventh grade is a time of strong contradictions: Although students feel a deeper sense of their own individuality, they are more self-conscious; although they are becoming increasingly confident, they also feel more insecure and are afraid of making mistakes; despite their interest in the world and other people, they are often overly self-absorbed.

The chemistry block addresses some of these contradictions. It strengthens students' sense of self by honing their thinking, which allows them to understand the causes of phenomena. It builds their confidence by giving them opportunities to work more actively and independently in the laboratory. And it reduces their self-absorption by requiring them to pay attention to the phenomena they are studying. Chemistry provides many opportunities for students to work in new ways, and the block is invariably one of the favorites of the year.

The emotional swings of the seventh grader are mirrored by the contrasts and opposites in the subjects they study: physical and chemical changes; fire and water; solid and gas; acid and base. As they study these subjects, students may have moments of self-perception, which form the foundation for self-knowledge.

In the last week of our chemistry block, we study solutions. A solution is composed of a solute (the dissolved substance) and a solvent (the liquid in which the solute is dissolved.) The solute in the solution is often in an ionic form, meaning that the compound is divided into positively and negatively charged molecules. Although the solute is usually invisible in a solution, it can be perceived by its effects: increase of mass, change in electrical conductivity, and taste. If a solution is saturated, it cannot hold any more solute ions. The ions combine, form a solid, and precipitate out of the solution.

We can view a chemical solution as a parable of nature. Think of the human being as a solute that is dissolved in the solution of spirit. When we are in the spiritual world between death and rebirth, the parts of our being are disparate, like ions with their positive and negative charges. To re-enter the physical world, we need to precipitate out of the spirit, to become whole again, to take on a physical form. Such thoughts should not be shared with the students, but pondering such a parable can help us recognize some of the cosmic mysteries that lie imbedded in the mundane details of everyday life.

In the practical realm, seventh graders are still receptive to demonstrations, but they will be most excited about performing lab activities themselves. Labs are a wonderful opportunity to teach students to work carefully and methodically and to treat the chemicals and processes they use with the respect they deserve. Seventh graders may not always seem outwardly responsive to what a teacher says, but they are affected deeply by what she does. The teacher's example of how to work with the equipment and materials will strengthen the students' relationship to matter and lay a moral foundation for their work in the material world.

Scheduling Considerations

If the chemistry block is three, rather than four, weeks, I suggest the following:

- Cover the topics of water and the generation of electricity during the physics block
- Do the first and second weeks of chemistry as recommended
- Work with the following topics during the third week:
  - *Day 1* – Limestone presentation; the lime cycle
  - *Day 2* – The lime cycle, continued; salt presentation; physical properties of salt
  - *Day 3* – Dissolution and precipitation of salt; synthesizing salt from acid and base
  - *Day 4* – Properties of acids and bases; effects of acids and bases on organic substances
  - *Day 5* – Testing for acids and bases

Week 1 of Grade 7 Chemistry

COMBUSTION

Most seventh graders look forward to their work in chemistry because it is unlike anything that they have done in school up to this point. As they anticipate the chemistry block, they invariably ask about the dangerous experiments they imagine doing. Though they may be disappointed by the response that nothing will explode, they will invariably be thoroughly engaged by the many different phenomena they will study.

In my experience, the best introduction to this first chemistry block is to learn the differences between physical and chemical changes. In the simplest terms, physical changes are changes of state and form, while chemical changes are changes of substance. The demonstrations of the first day not only help students recognize the hallmarks of chemical changes, they also give a preview of the subjects that will be considered more fully during the rest of the block: combustion, gases, and acids and bases.

Rudolf Steiner suggested that we begin the study of chemistry with the study of combustion. Why is this a good starting point?

Fire is one of humanity's earliest and most useful tools. It supported the development of culture by allowing people to heat buildings, cook food, fire clay, smelt and work with metals, and generate power. Fire is no less important in modern life. Almost all domestic, commercial, and industrial processes harness the transformative power of heat. In ancient cultures, fire was the medium of sacrifice to the gods, and we still use candles to grace special occasions and religious rituals.

Combustion is an archetypal chemical reaction. It requires fuel, oxygen, and heat—known as the fire triangle—and results in heat, light, and new substances. One can think of combustion as the opposite of photosynthesis, which is the basis for all life on earth. Through photosynthesis, plants assimilate carbon dioxide and water in the presence of sunlight to create glucose and oxygen. When plants burn, they release the energy of the sunlight they absorbed, and the substances they created from the glucose are transformed into carbon dioxide and water vapor.

In their study of combustion, students will experience the differences between the combustion of plant, animal, and mineral substances. One can think of the simplicity of plant combustion in connection to the etheric body, while the more complex (and odorous) combustion of animal substances reveals the residue of the animal's astral body. In contrast, the intense light and heat of the combustion of minerals express the purity of the purely physical body. Chemists learn about substances through qualitative and spectral analysis, but even on an elementary level, we can see how substances reveal their inner nature by how they burn.

Eugen Kolisko and many other teachers recommend that the study of combustion include phosphorus, coal, and sulfur, which serve as archetypes of combustion and are also found in our nerve-sense, rhythmic, and metabolic-limb systems. Because phosphorus is now a highly restricted substance and burning sulfur requires special venting, I have not included these topics. Instead, the "Botany of Flame" lab offers an opportunity to find correlations between the parts of the plant and the three-fold human being.

Having experienced the nature of combustion, students can now examine the fire triangle in more detail. Using fuel that is more refined results in more complete combustion, with no deposition of ash. Adding oxygen results in more intense and faster combustion, and increasing the heat or giving the combustible material greater access to the heat also enhances combustion.

The students' final experience for the week is to learn everything they can about a candle by observing it both lit and unlit. A candle contains a world of phenomena, as Michael Faraday showed in *The Chemical History of a Candle*. In these lectures, Faraday explored and explained the numerous aspects of chemistry encompassed by this common object. Because the candle burns with a controlled combustion process, students can explore the many aspects of combustion in a methodical manner.

This brings us to the end of the first week, and with this foundation, students will be ready to move on to the second week, when they will explore the gases of combustion.

TOPICS

- Physical and chemical changes
- Combustion of different types of substances
- Combustion of parts of the plant
- Enhancing combustion
- Controlled combustion

CHARACTERISTICS

- Physical and chemical changes have different attributes.
- Chemical changes can be revealed through a change of temperature.
- Chemical changes can be revealed by the production of gases.
- Chemical changes can be revealed by a change of color.
- Organic substances and some inorganic substances combust readily.
- The combustion of an organic substance reveals its source and/or composition.
- When parts of a plant burn, they reveal something about their relationship to the sun, air, water, and earth.
- Combustion can be enhanced by using refined fuels.
- Combustion can be enhanced by adding oxygen to the combustion process.
- Combustion can be enhanced by increasing the heat or the fuel's exposure to the heat.
- A candle burns with a controlled combustion process.

DAY	DEMONSTRATIONS AND ACTIVITIES	CHARACTERISTICS
1	■ Physical & Chemical Changes ■ Chemical Changes: Release of Heat ■ Chemical Changes: Release of Gas ■ Chemical Changes: Change of Color	■ Physical and chemical changes have different attributes. ■ Chemical changes can be revealed through a change of temperature. ■ Chemical changes can be revealed by the production of gases. ■ Chemical changes can be revealed by a change of color.
2	■ Combustion of Plant, Animal, & Mineral Substances	■ Combustion of an organic substance reveals its source and/or composition.
3	■ The Botany of Flame	■ When parts of a plant burn, they reveal something about their relationship to the sun, air, water, and earth.
4	■ Enhancing Combustion: Using Refined Fuel ■ Enhancing Combustion: Adding Oxygen ■ Enhancing Combustion: Increasing Fuel's Exposure to Heat	■ Combustion can be enhanced by using refined fuels. ■ Combustion can be enhanced by adding oxygen to the combustion process. ■ Combustion can be enhanced by increasing the heat or the fuel's exposure to the heat.
5	■ Controlled Combustion	■ A candle burns with a controlled combustion process.

DAY 1

In preparation for their work in chemistry, students should learn the differences between physical and chemical changes and become aware of the typical indicators that chemical changes have occurred.

DEMONSTRATION

1. Physical & Chemical Changes

CHARACTERISTIC: Physical and chemical changes have different attributes.

PROCEDURE

- **Combustion of Paper**
  - Rip a piece of paper into small pieces and put the pieces into a beaker that is about a quarter full of water.
  - Stir the paper and water.
  - Hold another piece of paper with a test tube clamp or set of tongs and ignite it.
  - Allow the paper to burn and the ashes to fall onto the sand.
- **Melting and Combustion of Paraffin**

  Melting and solidification of paraffin
  - Put the chafing dish fuel container in the sand tray.
  - Cut a small disk of paraffin off the bottom of a candle.
  - Cut it in half to remove the wick.
  - Light the chafing dish fuel.
  - Put the paraffin in a deflagration spoon and hold it over the flame until the paraffin melts.
  - Pour the melted paraffin into a beaker or bowl half full of water.

  Combustion of paraffin
  - Repeat, but this time allow the paraffin to ignite.

EQUIPMENT

medium beaker
stirring spoon
deflagration spoon
sand tray
tongs or test tube clamp
matches
butane lighter
spoon
small beaker or bowl
pliers

MATERIALS

piece of paper
water
white candle
chafing dish fuel

SAFETY

- A sand tray is an inexpensive and effective way to protect a table-top for any demonstration or lab activity involving combustion.

A sand tray can be made from a 9-inch x 9-inch square aluminum pan, half filled with sand.

- When burning the piece of paper, hold it with a test tube clamp or crucible tongs.
- To extinguish the chafing dish fuel, use a pair of pliers to handle the hot lid.

COMMENT

When you review on the following day, draw the students' attention to the following differences between physical and chemical changes:

- **Attributes**
  - *Physical change*: changes in the physical attributes of the substance, e.g., shape, size, appearance, texture, density, etc.
  - *Chemical change*: change in a substance's chemical properties or in its composition.
- **Reversibility**
  - *Physical change*: temporary and often reversible; original matter can be recovered by physical methods, e.g., filtration, evaporation, etc.
  - *Chemical change*: permanent and cannot be reversed; original matter has been changed and cannot be recovered.
- **Products**
  - *Physical change*: no new products are formed.
  - *Chemical change*: new products, whose properties are different from those of the reacting substances, are formed.
- **Energy**
  - *Physical change*: no or very little energy is absorbed or released.
  - *Chemical change*: energy is absorbed or released.

RELATED TOPICS

Examples of physical and chemical changes in nature and domestic and industrial settings

DEMONSTRATION

2. Chemical Changes: Release of Heat

CHARACTERISTIC: Chemical changes can be revealed through a change of temperature.

PROCEDURE

- Fill the beaker a quarter full with cold water.
- Hold the beaker steady a little above the table and have students feel the bottom of the beaker with the palm of their hands.
- Carefully pour a spoonful of sodium hydroxide into the water and stir.
- Be sure to add the sodium hydroxide to water, not water to the sodium hydroxide.
- Hold the beaker steady and have students feel the bottom of the beaker again.
- Have them feel the beaker a few minutes after the sodium hydroxide has been added.

SAFETY

- If holding the beaker is worrisome, it can be held by utility clamps attached to a laboratory stand.
- Sodium hydroxide is a powerful, caustic chemical, that needs to be used carefully.
- Everyone needs to wear safety goggles during this demonstration, and the teacher needs to wear gloves and an apron.
- Always add sodium hydroxide to water, ***not*** water to the sodium hydroxide.
- The sodium hydroxide solution can be diluted and safely flushed down the sink. (You can tell the students that sodium hydroxide, or lye, is often used to unclog sinks.)

EQUIPMENT

medium beaker

small spoon or laboratory scoop

MATERIALS

cold water

sodium hydroxide or lye

RELATED TOPICS

- Examples of exothermic and endothermic chemical reactions
- Chemical hand warmers
- Meals Ready to Eat (MREs)

DEMONSTRATION

3. Chemical Changes: Release of Gas

CHARACTERISTIC: Chemical changes can be revealed by the production of a gas.

EQUIPMENT

medium beaker
2 small spoons

MATERIALS

powdered citric acid
sodium bicarbonate (baking soda)
water

PROCEDURE

- Put a spoonful of citric acid in a beaker.
- Put a spoonful of baking soda in the same beaker and stir the two together.
- Pour a small amount of water into the beaker.

SAFETY

- Goggles and gloves should be worn for this demonstration.
- The baking soda and citric acid mixture can safely be flushed down the sink.

RELATED TOPICS

- Other common acid-base reactions that produce carbon dioxide

DEMONSTRATION

4. Chemical Changes: Change of Color

CHARACTERISTIC: Chemical changes can be revealed by a change of color.

PROCEDURE

- **Setup**

  Prepare indicator solution as described on page 99.
- **Process**
  - Fill the medium beaker half full of cabbage juice indicator solution. (For instructions on how to make cabbage juice indicator solution, go to p. 99)
  - Pour a spoonful of the powdered citric acid in the beaker a little at a time.
  - Observe the color change.
  - Fill the small beaker half full of bleach.
  - Use the dropper or pipette to add drops of bleach to the tinted indicator solution until the solution begins to turn transparent.
  - If you want the color change to be quicker, slowly pour the bleach into the indicator solution.

SAFETY

- When performing this demonstration, wear goggles, gloves, and an apron.
- Hold the beaker with the bleach steady while filling the pipette.
- The chemicals used in this demonstration can be diluted and safely flushed down the sink.
- The pipettes can safely be disposed of in the trash.

EQUIPMENT

medium beaker
small spoon
small beaker
dropper or pipette

MATERIALS

cabbage juice indicator solution
citric acid
bleach

DAY 2

If you have the time and the space, you can begin the study of combustion with a fire outside. The students will find it difficult to remain quiet for long, so don't plan on continuing to observe the fire until everything is fully combusted.

DEMONSTRATION

5. Combustion of Plant, Animal, & Mineral Substances

CHARACTERISTIC: Organic substances and some inorganic substances combust.

CHARACTERISTIC: The combustion of a substance reveals its source and/or composition.

EQUIPMENT

sand tray
butane lighter
Bunsen burner or propane torch
tongs or test tube clamp
aluminum pie or cake pan
window fan
pliers

MATERIALS

chafing dish fuel
dried small twig
dried leaf
hair
small feather
wool
magnesium ribbon

PROCEDURE

- Place the chafing dish fuel in the sand tray and light it.
- Using the tongs or test tube clamp, hold a twig over the flame. Once it ignites, move it away from the flame. Repeat with the leaf.
- Burn a small feather and small quantities of hair and wool.
- Tease apart a small amount of steel wool and try to ignite it.
- Extinguish the chafing dish fuel by replacing the lid with the pliers.
- Light the Bunsen burner or propane torch.
- Using a test tube clamp or tongs, hold one end of the magnesium ribbon in the flame until it ignites.
- While it burns, hold it over the pie or cake pan to catch the ash.

SAFETY

- Goggles, gloves, and an apron should be worn to perform this demonstration.
- Be aware that the animal substances burn with a pronounced odor, so use small quantities and use a window fan for ventilation.

- Because the magnesium burns with such an intense light, students should be instructed to look at it very briefly and then to look away.
- Magnesium also produces smoke, so ventilation is important.
- The cooled ash from the burned substances can be disposed of in the trash.

COMMENT

In conjunction with this demonstration, you may have the students read or listen to Jack London's short story "To Build a Fire."

RELATED TOPICS

Examples of combustion of other mineral substances

DAY 3

LAB ACTIVITY

6. The Botany of Flame

CHARACTERISTIC: When parts of the plant burn, they reveal something about their relationship to the sun, air, water, and earth.

This activity works best in groups of three students. Each group should have a tray with samples of the different parts of the plant. Provide each group with a table or chart to record observations. One student in each group can be assigned to serve as scribe and then her observations can be photocopied and shared.

Students should start by burning the root and then work their way up the plant by burning the stem, leaf, flower, and seed.

EQUIPMENT

sand tray
test tube clamp
chafing dish fuel
butane lighter or matches
pliers

MATERIALS

dried plant roots
dried twigs
dried leaves from two different types of trees
dried flowers such as hydrangeas
shelled almond, walnut, macadamia, or Brazil nuts

PROCEDURE

- Place the chafing dish fuel can in the sand tray and ignite it.
- Using the test tube clamp, hold each of the plant parts, one at a time, over the chafing dish fuel until it ignites, then move it aside from the flame to observe it burning on its own.

SAFETY

- Students can leave the chafing dish fuel burning while they take notes and prepare the next substances to burn, but they should remain aware of the open flame.
- The chafing dish fuel should be extinguished by replacing the lid with the pliers.

COMMENT

In the review, have the students be precise in their recollection of how different kinds of leaves or flowers burned, as well as the parts of the plant burned:

- *Root* – burns slowly, with little heat; much ash
- *Stem* – burns steadily, with fair amount of heat; less ash
- *Leaf* – burns quickly; very little ash
- *Blossom* – burns very quickly; almost no ash
- *Seed* – burns slowly with lots of heat and light; may not combust completely

RELATED TOPICS

- Types of plant-based fuels, e.g., wood, peat, coal, etc.
- Advantages and disadvantages of these fuels

DAY 4

DEMONSTRATION

7. Enhancing Combustion: Using Refined Fuel

CHARACTERISTIC: Combustion can be enhanced by using refined fuels.

All the demonstrations on this day should be done behind a laboratory safety shield.

PROCEDURE

- Put a watch glass in the sand tray.
- Using the pipette, drop a small amount of acetone onto a watch glass and ignite it.
- Repeat with drops of the alcohol or hand sanitizer on separate watch glasses.

SAFETY

- Wear goggles and gloves when performing this demonstration.
- Use small amounts of the substances; they will burn for long enough.
- Use a new pipette and a new watch glass for each substance.

EQUIPMENT

sand tray

plastic pipettes or droppers

watch glasses

MATERIALS

matches

acetone

rubbing alcohol or hand sanitizer

DEMONSTRATION

8. Enhancing Combustion: Adding Oxygen

CHARACTERISTIC: Combustion can be enhanced by adding oxygen to the combustion process.

Flash cotton and flash paper, which are available from magic stores or online sources, are two examples of substances that contain extra oxygen. This makes them highly combustible.

EQUIPMENT

laboratory safety shield
sand tray
test tube
laboratory stand with test tube clamp
Bunsen burner
tongs or test tube clamp
laboratory scoop
butane lighter

MATERIALS

flash cotton and flash paper
gummy bear candies
potassium chlorate

PROCEDURE

- Hold a piece of flash cotton with the tongs or a test tube clamp.
- Ignite the flash cotton with the butane lighter.
- Repeat with flash paper if desired. *Note: Use a different set of tongs or test tube clamp after each combustion. Otherwise, the flash cotton or the flash paper may ignite when it comes in contact with the hot holder.*
- Set up the laboratory stand with a clamp that holds the test tube upright, not slanted.
- Place the stand behind the laboratory safety shield.
- Place the Bunsen burner so that a gentle flame will be about 2 inches below the test tube.
- Using a clean scoop, put a small amount (about ¼ teaspoon) of potassium chlorate in the test tube.
- Put the gummy bear in the test tube on top of the potassium chlorate.
- Light the Bunsen burner and adjust the flame.
- When the potassium chlorate has melted and the gummy bear has reached its ignition temperature, the gummy bear will combust and there will be a rapid combustion with a large flame and lots of smoke.

SAFETY

- Potassium chlorate is a strong oxidizing agent and should be kept in safe chemical storage until it is used.
- A clean scoop needs to be used so that the potassium chlorate does not become contaminated.
- Before putting in the potassium chlorate, check the test tube to make sure that there are no cracks or chips.
- The test tube with the burned sugar may need to be soaked for a while before it can be cleaned.

COMMENT

Make sure that flash paper and flash cotton are stored safely when not in use.

RELATED TOPICS

- Sugar's use in military mines
- Other oxidizers, such as potassium nitrate
- Gunpowder

DEMONSTRATION

9. Enhancing Combustion: Increasing the Fuel's Exposure to Heat

CHARACTERISTIC: Combustion can be enhanced by increasing the heat or the fuel's exposure to the heat.

This demonstration allows a finely dispersed powder to come in contact with a flame. Lycopodium powder (the spores of club mosses) is ideal, but cornstarch or non-dairy creamer can also be used.

Although this demonstration is very dramatic, it is not dangerous if students are standing well back from the demonstration area.

EQUIPMENT

funnel

plastic tubing that fits on the end of the funnel, 2′ long

small piece of wire screen

small piece of tissue paper

Bunsen burner

spoon or laboratory scoop

MATERIALS

butane lighter or lab striker

lycopodium powder

PROCEDURE

- Attach the plastic tubing to the funnel.
- Put a small piece of wire screen and a small piece of tissue paper in the narrow bottom of the funnel so that the powder cannot fall into the tubing.
- Light the Bunsen burner and adjust the flame so that it is a high yellowish flame.
- Carefully pour about a teaspoon of lycopodium powder onto the tissue paper.
- Kneel down so that you are below the barrel of the Bunsen burner but can ***hold the funnel upright*** next to the top of the Bunsen burner's barrel.
- Hold the free end of the plastic tubing and bring it to your lips.
- With a quick, hard exhalation, blow the powder into the flame.
- As soon as the powder has combusted, turn off the Bunsen burner.

SAFETY

- Make sure that the students are well back from the flame column.
- The piece of tissue paper may ignite and float to the ground. Be prepared to extinguish it.
- To clean up, disconnect the funnel from the hose and wipe it. Blow through the hose to clear away any remaining powder.

COMMENT

Because the ignition of the lycopodium powder is so dramatic, students will probably insist on seeing the demonstration repeated. I do not recommend doing it more than twice, though.

During the review, students should be told about explosions in coal mines, grain elevators, or cotton mills. You may wish to show a video of a dramatic fine-particle explosion.

DAY 5

LAB ACTIVITY

10. Controlled Combustion: The Candle

CHARACTERISTIC: A candle burns with a controlled combustion process.

PROCEDURE

- Each student should do the following.
- Observe the candle before it is lit.
- Light the candle and make as many observations as possible.
- Rub a little bit of soot from an extinguished match so that small particles fall into the melted wax in the well of the candle.
- While the candle is burning, hold the back of the spoon in the flame.
- Next, blow out the candle and observe the smoke.
- Light both candles and let them burn for about a minute.
- Using the second candle as an ignition source, light a match.
- Blow out the first candle and hold the lit match about 1 inch away in the smoke from the extinguished candle. If it is timed correctly, the flame from the match will jump through the unburned vapors and relight the candle.

EQUIPMENT

enough spoons for each student

MATERIALS

two white paraffin candles per student

matches for each student

COMMENT

Because combustion is often such a rapid process, it is hard to observe its subtler aspects. A burning candle provides the opportunity to examine the combustion process carefully.

Observations should include:

- physical attributes of the candle and wick
- process of lighting the candle
- zones of the flame

- changes in the wick
- well of molten wax on top of candle
- movement of soot particles in candle's well
- deposition of soot on the back of a spoon held over the flame
- dripping of wax
- process of extinguishing candle and the release of smoke
- re-lighting the candle's unburned vapors

RELATED TOPICS

- Other examples of domestic and industrial controlled combustion processes
- Candle-making through history
- Animal and vegetable sources of candle wax

Week 2 of Grade 7 Chemistry

GASES OF COMBUSTION

When we observe combustion, we most easily experience the flames, heat, smoke, and ash. But combustion also has some less visible aspects, specifically the consumption of oxygen and the production of water vapor and carbon dioxide. During the second week of the block, we will focus on the characteristics of these invisible aspects of combustion.

Put a lit candle under a jar and observe what happens: slight condensation forms on the inside of the jar; the flame begins to gutter and becomes smaller; and finally, the candle goes out and expels a tendril of smoke. What was happening in the jar that caused the candle to go out?

Because most students have already learned that fire needs oxygen, they will say that the candle went out because it used up all the oxygen in the jar. This is true, but it is not the whole truth. The whole truth is much more complicated. In addition to consuming oxygen, the candle is producing carbon dioxide. This carbon dioxide absorbs some of the candle's heat, lowering the temperature around the wick, thereby slowing down the conversion of the paraffin from its liquid to gaseous state. It also pushes oxygen away from the wick so that the candle does not have enough oxygen to continue to burn. Eventually, the atmosphere around the candle flame does not contain enough oxygen to support combustion, but there is still plenty of oxygen in the jar.

These details need not be shared with students, but it is helpful for the teacher to know how complicated chemical reactions can be. Based on their observations, students will agree that something changed inside the jar that made the candle stop burning. Was something taken out of the air? Was something added to the air? Students can be told that both are true, and that these "somethings" will be the focus of the week's work.

Students may also be told that carbon dioxide is a relatively recent name for a gas that has been known for centuries. The modern name is based on its chemical formula, but long before scientists understood the atomic nature of matter, this gas had been discovered and its properties studied.

One of the first scientists to discover the gas was Jan Baptist von Helmont (1580–1644), who discovered that burning coal produced an "air" that stifled the combustion of other substances. Joseph Black (1728–1799), a British scientist, discovered the same gas by heating various carbonates intensely and then measuring their loss of weight. Black also immersed carbonates in acid solutions and discovered that the carbonates released a gas. He called the gas "fixed air" because it had been fixed in the carbonate until released by the acid. Black described the parts of the lime cycle, and he developed the limewater test for carbon dioxide that we still use today. A contemporary of Black's, Joseph Priestley (1733–1804), also worked with this gas, producing it by reacting an acid and a base. Priestley discovered that this gas was soluble, and he developed a commercial process for producing artificially carbonated water.

Teachers may choose to share brief biographies of these scientists with their students, or they may decide to treat them as biographical background information. What is important is that the students discover and observe the properties of carbon dioxide themselves and get to know this unusual gas.

Students can produce carbon dioxide by mixing baking soda (a carbonate) and vinegar (an acid). The gas will be revealed by the mass of bubbles that arise, and the fact that the gas does not support combustion can be demonstrated. The weight of carbon dioxide can then be shown in a dramatic demonstration in which it is poured onto a candle and extinguishes the flame.

To test for the presence of carbon dioxide, students can follow Joseph Black's method of bubbling the gas through limewater and precipitating lime out of the solution. This method can also allow them to confirm that their exhaled breath contains carbon dioxide.

For a final experience of carbon dioxide, the class can work with carbon dioxide in its solid form, dry ice—a substance with remarkable properties.

From all of these demonstrations and activities, students will get to know many of carbon dioxide's physical and chemical properties. In eighth grade, they will learn the gas's role in photosynthesis and will appreciate how plants transform the carbon from the atmosphere into living substance.

During the second half of this week, we study oxygen. Here again Joseph Priestley figures prominently, for he was the first scientist to isolate oxygen and examine its properties. Priestley used a lens to heat mercuric oxide in a bell jar, but we will use a safer substance—hydrogen peroxide—and will liberate the oxygen from the compound by using yeast as a catalyst.

Students will learn how to capture oxygen "over water," and then they will see how various substances combust in pure oxygen. As dramatic as it may be to see carbon, steel wool, and magnesium burn in oxygen, the demonstration with the two candles—one burning in a jar of air and the other in a jar of oxygen—is ultimately more instructive. The students' observations of how long and how intensely the candle burns in oxygen, how much of the candle is consumed, and how much water vapor is produced will help them appreciate the benefits of the relatively low percentage of oxygen in the air around us.

Burning substances in oxygen is a wonderful way to end the week, but if teachers want an interesting reprise, they can include two final demonstrations the next week: the synthesis of carbon dioxide by burning carbon in oxygen and the combustion of magnesium in carbon dioxide. Both of these demonstrations are more appropriate for high school, but they give the students an appreciation for the mysteries of chemical reactions and whets their appetite for future studies in chemistry.

TOPICS

- Byproducts of combustion
- Production of carbon dioxide
- Properties of carbon dioxide
- Testing for carbon dioxide
- Production of oxygen
- Combustion in oxygen
- Synthesis and decomposition of carbon dioxide

CHARACTERISTICS

- The rate or extent of combustion may be reduced if it occurs in a container or enclosed space.
- The combustion of organic compounds produces light, heat, water

vapor, and carbon dioxide.

- Carbon dioxide is released by the reaction of a liquid acid with a carbonate base.
- Carbon dioxide stifles the combustion of most substances.
- Carbon dioxide causes the precipitation of lime out of a limewater solution.
- Carbon dioxide is heavier than air.
- Carbon dioxide can be poured from a container.
- Carbon dioxide can exist in solid and gaseous states.
- Solid carbon dioxide sublimates to gas at normal temperatures.
- Gaseous oxygen is a byproduct of hydrogen peroxide catalyzed by yeast.
- Gases can be collected using the water displacement method.
- Extra oxygen enhances combustion.
- Carbon dioxide can be synthesized by burning carbon in oxygen.
- Magnesium can combust in carbon dioxide by utilizing the oxygen in the gas.

DAY	DEMONSTRATIONS AND ACTIVITIES	CHARACTERISTICS
1	■ Products of Combustion	■ The rate or extent of combustion may be reduced if it occurs in a container or enclosed space. ■ The combustion of organic compounds produces light, heat, water vapor, and carbon dioxide.
2	■ Carbon Dioxide Does Not Support Combustion ■ Testing for Carbon Dioxide with Limewater ■ Testing Breath for Carbon Dioxide ■ Properties of Carbon Dioxide's Weight	■ Carbon dioxide is released by the reaction of a liquid acid with a carbonate base. ■ Carbon dioxide stifles the combustion of most substances. ■ Carbon dioxide causes the precipitation of lime out of a limewater solution. ■ Carbon dioxide is heavier than air ■ Carbon dioxide can be poured from a container.
3	■ Demonstrations with Dry Ice	■ Carbon dioxide can exist in solid and gaseous states. ■ Solid carbon dioxide sublimates to gas at normal temperatures.
4	■ Generating Oxygen ■ Capturing Oxygen by Water Displacement	■ Gaseous oxygen is a byproduct of hydrogen peroxide catalyzed by yeast. ■ Gases can be collected using the water displacement method.
5	■ Combustion in Oxygen	■ Extra oxygen enhances combustion.
6	■ The Synthesis of Carbon Dioxide ■ The Combustion of Magnesium in Carbon Dioxide	■ Carbon dioxide can be synthesized by burning carbon in oxygen. ■ Magnesium can combust in carbon dioxide by utilizing the oxygen in the gas.

DAY 1

LAB ACTIVITY

1. Products of Combustion

CHARACTERISTIC: The rate or extent of combustion may be reduced if it occurs in a container or enclosed space.

CHARACTERISTIC: The combustion of organic compounds produces light, heat, water vapor, and carbon dioxide.

It is fitting to begin a week that includes many complex phenomena with a seemingly simple one.

EQUIPMENT

mason jars, quart-sized

MATERIALS

2 candles per student
matches

PROCEDURE

- Each student is given two candles and one mason jar.
- Set both candles upright on the table top, about 9 inches apart, and light them.
- Put the mason jar over one of the candles and observe until the candle is extinguished.
- Lift the jar and quickly place it over the second burning candle.
- Observe what happens to that candle.

COMMENT

During the review, challenge the students to avoid referring to oxygen or carbon dioxide as a way of explaining the phenomenon. Try to lead them to the recognition that it is clear that the conditions supporting combustion change as the candle burns. This could mean that the candle is using up something required for combustion or that it is producing something that inhibits combustion. You can then mention that these "somethings" are oxygen (used up during combustion) and carbon dioxide (produced by combustion).

Note: Other teachers' guides may recommend a demonstration that I advise against. The demonstration involves placing a candle in a shallow pan of water and putting a jar over it. As the candle burns, and when it is extinguished, water rises into the jar. The conventional explanation is that as the candle utilizes the oxygen in the air in the jar, water fills up that space. I recommend not doing this demonstration for two reasons:

First, the conventional explanation doesn't take into account that as oxygen is depleted, carbon dioxide is produced, so the candle's consumption of oxygen does not create a vacuum.

Second, the heat of the combustion process causes the air in the jar to expand. If you observe carefully, you can see small bubbles being forced out of the bottom of the jar. When the air in the jar cools, a partial vacuum is created, and the air pressure outside the jar forces water into the jar.

DEMONSTRATION

2. Carbon Dioxide Does Not Support Combustion

CHARACTERISTIC: Carbon dioxide is released by the reaction of a liquid acid with a carbonate base.

CHARACTERISTIC: Carbon dioxide stifles the combustion of most substances.

PROCEDURE #1

- Put several tablespoons of baking soda into the beaker.
- Pour in about ¼ cup of vinegar.
- When the bubbling has subsided, light a match and slowly lower it into the beaker. It will be extinguished well above the level of the liquid.

EQUIPMENT

beaker, medium or large

spoon

MATERIALS

short candle supported by a piece of wire

baking soda

vinegar

matches

PROCEDURE #2

- Light the candle on the wire and slowly lower it into the beaker.
- As the candle starts to go out, raise it a little bit so that it doesn't go out.
- Repeat several times to show the level of carbon dioxide in the beaker.

RELATED TOPICS

- Naturally carbonated springs
- The soft-drink industry
- Industrial and commercial uses of carbon dioxide

DAY 2

DEMONSTRATION

3. Testing for Carbon Dioxide with Limewater

CHARACTERISTIC: Carbon dioxide causes the precipitation of lime out of a limewater solution.

PROCEDURE

- **Setup**
  - Create lime milk prior to this demonstration day by adding several teaspoons of calcium oxide or calcium hydroxide to about 250 ml of water in a flask. (Proportions are not important.)
  - Stir the mixture for about 5 seconds and filter it to create limewater.
  - Stopper the flask of limewater to prevent a layer of lime from forming on the top.

- **Process**
  - Fill a small beaker with about 1 inch of the limewater that has been prepared in advance. If this limewater turns milky, it will reveal the presence of carbon dioxide.
  - Set up a flask as follows: Use a stopper that has a short piece of glass tubing attached to it and a plastic tube attached to the glass tubing. The plastic tube should be long enough to extend into the beaker of limewater.
  - Put baking soda in the flask and pour in some vinegar to create carbon dioxide. Stopper the flask.
  - Put the plastic tube into the beaker of limewater and bubble the carbon dioxide through the limewater.
  - The limewater should turn milky with precipitated lime.
  - In addition to or as an alternative to the demonstration described above, you can put an effervescent antacid tablet into water in the flask. The tablet produces carbon dioxide, which will turn the limewater in the beaker milky.

EQUIPMENT

two 250 ml Erlenmeyer flasks
funnel
stopper to fit flask
short piece of glass tubing
plastic tubing
small beaker

MATERIALS

limewater
filter paper
baking soda
vinegar
effervescent antacid tablets

SAFETY

- Teachers should wear goggles and gloves when handling calcium oxide, calcium hydroxide, or limewater.
- Calcium oxide and calcium hydroxide can be neutralized with vinegar and be safely flushed down the sink.

RELATED TOPICS

- Acid indigestion
- Antacids and how they work

ACTIVITY

4. Testing Breath for Carbon Dioxide

CHARACTERISTIC: Human exhalation contains carbon dioxide.

EQUIPMENT

test tube rack

test tubes, 1 per student

MATERIALS

limewater

drinking straws, 1 per student

PROCEDURE

- Pour about 1 inch of limewater into each test tube and put a straw in the test tube.
- Have students breathe gently into the limewater until it turns milky.

COMMENT

If students continue to breathe into the limewater after the lime precipitates, they may find that the limewater begins to clear again. This is because the carbon dioxide in their breath turns the water acidic, which dissolves the precipitated lime.

SAFETY

- Students should wear goggles and gloves for this activity.
- They should be told to breathe gently into the limewater so that it doesn't spurt out of the test tube.
- Limewater can be diluted, neutralized with vinegar, and safely flushed down the sink.

ACTIVITY

5. Properties of Carbon Dioxide's Weight

CHARACTERISTIC: Carbon dioxide is heavier than air.

CHARACTERISTIC: Carbon dioxide can be poured from a container.

EQUIPMENT

Erlenmeyer flask, 500 or 1000 ml

two small jars or beakers

MATERIALS

baking soda

vinegar

two tea lights

matches

PROCEDURE

- Create carbon dioxide in the large flask using baking soda and vinegar, as described earlier.
- Use a match to test that the carbon dioxide has filled the flask. If so, the match will go out as soon as you lower it into the flask.
- Place tea light candles in two small jars or beakers and light them.
- Tilt the flask gently over one of the tea lights and pour the gas onto the flame. The flame should go out.
- Repeat with the second tea light.

SAFETY

- Teachers should wear goggles when performing this demonstration.
- The baking soda-vinegar mixture can safely be flushed down the sink.

COMMENT

Students will enjoy hearing about naturally occurring carbon dioxide in springs and about the Cave of Dogs near Naples, which contains a fumarole that expels carbon dioxide. Because the carbon dioxide is heavier than air, it lingers near the floor of the cave and will asphyxiate any animals who aren't tall enough to walk above the level of the gas.

RELATED TOPICS

- Natural sources of carbon dioxide
- How carbon dioxide is used to extinguish fires

DAY 3

Carbon dioxide has the remarkable property that it exists in a solid and gaseous form but not as a liquid. Liquid carbon dioxide can only be created at high pressure and very low temperatures.

Solid carbon dioxide is called dry ice. It can typically be obtained in a supermarket. Dry ice is extremely cold and can cause frostbite, so it needs to be handled with gloves or tongs. It can only be stored in a well-insulated cooler, and even then, it will sublimate in a few hours. Be sure to inform yourself fully about working with dry ice before attempting these demonstrations.

DEMONSTRATION

6. Demonstrations with Dry Ice

CHARACTERISTIC: Carbon dioxide exists in a solid and gaseous state.

CHARACTERISTIC: Solid carbon dioxide sublimates to gas at normal temperatures.

PROCEDURE

Here are five brief demonstrations:

1. Break a piece of dry ice into smaller pieces and have the students observe how it sublimates (goes from solid to gas).
2. Put a few small pieces of dry ice in a beaker and test for the presence of carbon dioxide with a lit match or by lowering a lit candle on a wire. The flame will go out if carbon dioxide is present.
3. - Fill a clean plastic 1-liter soft-drink bottle almost full with cold distilled water.
   - Drop in a piece or a few small pieces of dry ice and cap tightly.
   - Place the bottle behind the laboratory safety shield or well away from students until the reaction subsides. *Note: I have never had a bottle explode but have been warned that it can happen.*

- When the bubbling has stopped, pour out a little bit of the carbonated water into cups for students to taste.
- Test the carbonated water for carbon dioxide by pouring a small amount into a beaker and adding some limewater. If the limewater turns milky, carbon dioxide is present.

4. - Put a number of chunks of dry ice into a large bowl or aquarium.
   - Pour warm water over the carbon dioxide. *Note: The clouds that form are not carbon dioxide but condensed water vapor.*
   - Blow soap bubbles with a wand so that they fall onto the clouds of water vapor. They should bob on the "lake" of carbon dioxide.
5. - Put a small chunk of carbon dioxide into each student's cup.
   - Have students add water to the carbon dioxide and observe how it sublimates.
   - Students can breathe the condensing water vapor and/or taste the slightly carbonated water.

SAFETY

- The teacher should wear protective gloves to handle the dry ice and only pick it up with gloves or a pair of pliers or tongs.
- Students should not handle the dry ice with bare hands.
- The teacher should wear goggles and gloves when working with limewater.
- Students may wish to taste the carbonated water. You will have to decide whether to allow this, because there is no assurance that the dry ice was uncontaminated.
- The remaining dry ice can be allowed to sublimate, and the water can be poured down the sink.
- The limewater can be diluted, neutralized with vinegar, and safely flushed down the sink.

RELATED TOPICS

- Commercial uses of dry ice
- How commercial soft drinks and sparkling water are carbonated
- Home water carbonators

EQUIPMENT

laboratory safety shield
gloves
tongs or pliers
medium or large beaker
empty plastic 1-liter soft-drink bottle with cap
large bowl or aquarium

MATERIALS

3–5 lbs. of dry ice stored in an insulated cooler
matches
short candle on wire
distilled water
limewater
water
soap bubble solution
bubble wand
small paper cups
clear plastic cups, 1 per student

DAY 4

Students have already experienced enhanced combustion with the oxygen-rich flash cotton or flash paper and with the release of oxygen from potassium chlorate. In this lab, they will learn how to generate oxygen themselves.

LAB ACTIVITY

7. Generating Oxygen

CHARACTERISTIC: Gaseous oxygen is a byproduct of hydrogen peroxide catalyzed by yeast.

EQUIPMENT

test tubes, 1 per student

test tube racks or beakers to hold the test tubes

laboratory scoops, 1 per each pair of students

MATERIALS

baking yeast

fresh hydrogen peroxide

small paper baking cups

candles

matches

small wooden splints or coffee stirrers

PROCEDURE

- Students should work individually on this lab. Have them carry out the following steps.
- Fill a test tube with 1 inch of hydrogen peroxide.
- Light a candle and have it ready to light the splints.
- Pour a small amount of yeast (pea-sized portion) into the hydrogen peroxide.
- Wait for the bubbling to stop.
- Light the splint with the candle.
- After it has burned for a few seconds, blow out the splint. Its tip should still glow.
- Put the glowing end of the splint into the test tube. Do not put it into the liquid at the bottom of the test tube.
- The glowing end of the splint should burst into flame in the presence of oxygen.
- Quickly take out the splint (so it doesn't produce too much carbon dioxide), blow out the flame, and try to re-ignite it.

SAFETY

- Students should wear goggles and gloves for this activity.
- They should exercise caution and remember that the candle for

lighting splints will remain lit on their desk.

- The hydrogen peroxide-yeast mixture can be diluted and safely flushed down the sink.

COMMENT

This activity can be done on a larger scale by using beakers and hydrogen peroxide of a higher concentration—30% or 40% hydrogen peroxide is available in beauty-product stores that sell hair-dying products.

The use of hydrogen peroxide for cleansing wounds can be examined. Students may notice that blood foams when it comes in contact with hydrogen peroxide. The little bubbles are bubbles of oxygen.

RELATED TOPICS

- The use of manganese dioxide as a catalyst to produce oxygen
- The use of hydrogen peroxide as a bleaching agent for hair, teeth, etc.
- How oxygen is produced on a commercial scale
- The use of oxygen in medicine

DEMONSTRATION

8. Capturing Oxygen

CHARACTERISTIC: Gases can be collected using the water displacement method.

PROCEDURE

- Cover the bottom of a plastic container with about 2 inches of water.
- Fill a quart mason jar with water and lightly cap it.
- Place the jar upside down in the water and remove the cap.
- Connect the oxygen canister to the oxygen port of the regulator. (The MAPP gas cylinder will not be used.)

EQUIPMENT

a shoebox-sized plastic container

7 mason jars with lids

MAPP gas/oxygen canisters with regulator

plastic tubing to attach to regulator

- Attach the plastic tubing to the end of the nozzle of the regulator hose.
- Set up the oxygen canister with the tubing so that the tubing can easily reach into the plastic container.
- Place the tubing in the water and turn on the regulator attached to the oxygen canister so that bubbles of oxygen come out of the hose.
- Hold the end of the tubing underneath the jar to allow the bubbles to force out the water. As the jar fills with oxygen, the water is forced out.
- When the jar is full of oxygen, turn off the flow of gas and cap the jar before removing it from the water.
- Repeat to fill 7 jars of oxygen. In the process, you may need to empty the plastic container as it fills with water from the jars.
- Put the jars of oxygen in a safe place. They will be used for demonstrations on Day 5.

SAFETY

- Make sure that the canister of oxygen is stable throughout this demonstration. It may require someone to hold it while the gas is being captured.
- There should be no open flames while oxygen is being transferred.

COMMENTS

- MAPP gas/oxygen canisters are sold in hardware stores. For this demonstration, the MAPP gas will not be used. Students should be told about the pneumatic trough, a piece of equipment especially designed to capture gas by displacement.

RELATED TOPICS

- Other ways of capturing gases
- MAPP gas and its uses

DAY 5

As the culmination of the week on gases associated with combustion, we will burn various substances in pure oxygen. Students will note that the combustion process is heightened in every respect: more light, more heat, faster reaction speed, and greater amount of byproducts.

DEMONSTRATION

9. Combustion in Oxygen

CHARACTERISTIC: Extra oxygen enhances combustion.

PROCEDURE

- **Setup**
  - Prepare the equipment and the substances to be lowered into the jars of oxygen as follows:
  - Drill small holes through the 4 mason jar lids.
  - Cut pieces of brazing rod, insulation rod, or stiff straight wire (e.g., pieces of coat hanger) into 10-inch lengths. You will need a total of six pieces.
  - Prepare these substances to be used in the demonstration:
    small piece of lightly teased steel wool
    drawing charcoal, 4–5 inches long
    piece of magnesium ribbon, 2–3 inches long
  - Using thin wire, attach the substances to ends of the rods. You will need two rods for each of the substances.
- **Process**
  - Arrange two candles slightly apart so that they can be covered by mason jars.
  - Unscrew the lid from one jar of oxygen, but do not remove it. Leave the lid lightly on top of the jar until it is ready to be used.
  - Light the candles.
  - Put an empty jar over one candle. Remove the lid from the jar full of oxygen and put it over the other candle.
  - Observe and compare the candles' combustion.

EQUIPMENT

laboratory safety shield
5 mason jars filled with oxygen
4 mason jar lids with holes
2 empty mason jars
Bunsen burner
drill with small bit

MATERIALS

2 candles
matches
brazing rod, insulation rod, or hanger wire, 4 pieces cut to a length of 10″
small piece of lightly teased steel wool
drawing charcoal, 4–5″ long
piece of magnesium ribbon, 2–3″ long
thin wire

Burning substances in oxygen

For each of the substances (charcoal, steel wool, and magnesium ribbon) to be burned, you will first light a sample of the substance held by a pair of tongs and observe how it burns in the air. Then you will ignite the sample of the substance that is attached to a brazing rod with the thin wire and lower it into a jar of oxygen. Wrap some tape around the brazing rod above and below the lid so that the rod doesn't slip and the substance doesn't touch the bottom of the jar.

- *Drawing Charcoal*
  - Put the brazing rod with the drawing charcoal through one of the lids with a hole in it.
  - Prepare the jar of oxygen by unscrewing the lid and leaving it loosely on top of the jar.
  - Hold a sample of the charcoal in a pair of tongs and observe how it burns in the air.
  - Then light the charcoal attached to the brazing rod inserted through a lid and lower the rod into a jar of oxygen.
  - Repeat if desired.
- *Steel Wool*
  - Prepare a jar of oxygen by having its lid ready to remove.
  - Hold a piece of steel wool in a pair of tongs or with a test tube clamp.
  - Ignite it and observe how it burns in the air.
  - Then light a sample of the steel wool that has been attached to the brazing rod inserted through a lid and lower the rod into a jar of oxygen.
  - Repeat if desired.
  - Note: *The steel will burn at a high temperature, shooting off sparks and small drops of molten steel. They may crack the jar, but it should not break.*
  - Students may enjoy this so much that they will ask for it to be repeated. That is why you need to have two brazing rods and lids with steel wool.

- ***Magnesium Ribbon***
- Prepare a jar of oxygen by having its lid ready to remove.
- Light the propane torch or Bunsen burner.
- Holding a piece of magnesium ribbon with tongs or a test tube clamp, light the end of the ribbon and observe how it burns in the air.
- Then light a sample of the magnesium ribbon that has been attached to a brazing rod inserted through a lid and lower it into a jar of oxygen.
- *Note: I suggest not burning another piece of magnesium ribbon in oxygen; repeating the procedure may diminish the powerful impression that the first combustion has made.*

SAFETY

- Teachers should wear goggles, gloves, and an apron to perform this demonstration.
- These demonstrations should be performed behind a laboratory safety shield.
- Students should be reminded not to look at the burning magnesium for more than a very brief moment at a time.
- If a jar cracks during the demonstration, it should be disposed of in the trash.

COMMENT

During the review, ask students to consider the implications for having an atmosphere that is richer in oxygen than ours is (currently about 21 percent).

DAY 6

The next two demonstrations can be performed during the review on the following Monday.

Introduce these demonstrations by telling the students that it is possible to synthesize compounds from elements and to decompose compounds into their component elements.

DEMONSTRATION

10. Synthesizing Carbon Dioxide

CHARACTERISTIC: Carbon dioxide can be synthesized by burning carbon in oxygen.

EQUIPMENT

Bunsen burner
matches or lab striker

MATERIALS

mason jar filled with oxygen
mason jar lid with hole
brazing rod or wire hanger, 10″ long
drawing charcoal
thin wire
limewater

PROCEDURE

- Prepare a lid with a brazing rod and drawing charcoal as described for the previous demonstration.
- Pour a little limewater into a jar filled with oxygen.
- Light the charcoal and put it in the jar filled with oxygen and limewater.
- When the combustion has ended, shake the jar and see how the limewater turns milky.
- You have demonstrated that you have synthesized carbon dioxide from carbon and oxygen.

SAFETY

- Teachers should wear goggles and gloves when working with limewater.

DEMONSTRATION

11. Decomposing Carbon Dioxide

CHARACTERISTIC: Magnesium can combust in carbon dioxide by utilizing the oxygen in the gas.

Compounds can also be decomposed through combustion.

MATERIALS

mason jar, quart-sized
mason jar lid with hole
baking soda
vinegar
magnesium ribbon
brazing rod or wire hanger, 10″ long
thin wire

PROCEDURE

- Prepare a lid with a brazing rod that has a piece of magnesium ribbon attached with the thin wire.
- Prepare a mason jar of carbon dioxide by pouring ¼ cup of vinegar onto approximately 3 tablespoons of baking soda.
- Test with a match to see if the jar is filled with carbon dioxide.
- Light the magnesium ribbon with the Bunsen burner or propane torch and lower it into the jar.
- Students will notice that, as the magnesium burns, tiny flecks of carbon are produced.
- The magnesium bonds with the oxygen and leaves the carbon behind.

SAFETY

- The jar in which the magnesium burns may crack. If it does, dispose of it in the trash.
- The limewater can be diluted, neutralized with vinegar, and safely flushed down the sink.

Week 3 of Grade 7 Chemistry

WATER & THE LIME CYCLE

Although our planet is called "Earth," it could more appropriately be called "Water," because over 70 percent of the earth's surface is covered by water, and water is also present underground and in the atmosphere. Of all the water on earth, only a tiny fraction is fresh water, and most of that makes up the polar ice caps. Water is the only substances on earth that exists in all three states in the earth's temperature range, and because water absorbs and releases heat so slowly, it helps the earth maintain a temperature that supports life.

The water cycle is a vast sunlight-driven atmospheric engine that moves millions of tons of water every year. Water on the earth's surface evaporates and is carried into the atmosphere, where it cools, condenses, and eventually precipitates. Most of that precipitation falls on oceans or other bodies of water. Some of the precipitation that falls on land is absorbed by living organisms, who respire or transpire water vapor back into the atmosphere, and some of it flows into underground streams and reservoirs. Surface water flows into streams, rivers, and lakes, and some of it eventually returns to the oceans.

Either in their study of heat in physics or during this chemistry block, students need to learn about water's changes of state and should be able to name all of the different processes: melting/freezing, evaporation/condensation, and sublimation/deposition. Students should recognize that all of these changes are a result of heat gain or loss and that heating and cooling cause other substances besides water to go through similar changes.

Students will then have the opportunity to measure the boiling and freezing points of water. Because water is so readily available and changes state easily and consistently, it proved to be an ideal substance for establishing temperature scales. Students should learn about the development of the Fahrenheit and Celsius scales, and they should be introduced to the Kelvin scale, which will be used in their more advanced science studies.

Having established that pure water boils at 100° Celsius and freezes at 0° Celsius (at normal atmospheric pressure), students can then try raising the boiling point and lowering the freezing point by adding salt to the water. This lab activity sets the stage for working with salt solutions during the next week, and it is a good time to consider water's role as a universal solvent.

Adding a solute—a substance that dissolves in a solvent—also affects water's electrical conductivity. Students will find that distilled water does not conduct electricity, but as more solute is added, it becomes increasingly conductive. The work with water can culminate by using electricity to decompose water into its component gasses: hydrogen and oxygen. Although the electrolysis of water could wait for high school, it makes a strong impression on the young adolescents who may casually refer to water as H_2O but not grasp what the formula really means.

The study of water is not nearly as dramatic or captivating as the study of combustion or gases, but it is nevertheless important for students to come to know, understand, and appreciate this amazing substance that affects so many aspects of our world and of our lives.

The last two days of the week move into a new realm: salts. I recommend that we begin with limestone because it has an organic origin, comes in many forms, and is more complex than common salt. When students study sodium chloride in the following week, they will observe some of the properties of salt that are not as easily perceptible as the properties of limestone.

Calcium is an essential nutrient, used by plants in their cell walls and by animals in their shells and bones. In nature, calcium, in the form of lime, is continually being deposited and dissolved. Limestone is always formed in water. Some types are created from the precipitation of calcium carbonate from dissolved skeletons of marine organisms; others are made up of sediments that include the skeletons themselves. Limestone can be metamorphosed into marble, which sometimes reveals its organic origin by the forms of the organisms in the stone.

In their study of geology, students may have learned about the typical limestone landscape—gentle with rolling hills, well-watered, and verdant. Sometimes the limestone bedrock is exposed and fossils may be found. Often there are waterfalls or hollows and caves carved out of the limestone by slightly acidic water. Students may also have seen—or seen pictures of—chalk cliffs or marble quarries, which are other examples of limestone. It is essential that we connect the substances that we examine in class with their presence in nature so that students have the largest possible context for what they are learning.

Lime's chemical name is calcium carbonate, and it is made up of calcium, carbon, and oxygen. When we heat lime (calcium carbonate), carbon dioxide is driven off, and what remains is quicklime, more formally called calcium oxide. When quicklime comes into contact with water, it produces slaked lime, whose chemical name is calcium hydroxide. When slaked lime comes into contact with carbon dioxide, it forms lime again, and water is driven off. Although we don't need to share the formula with the students, it is good for the teacher to know what is happening in chemical terms in each of the three stages of the limestone cycle.

1. *in common language:*

   lime + heat = quicklime + carbon dioxide

   in scientific language:

   calcium carbonate + heat = calcium oxide + carbon dioxide

   the formula:

   $$CaCO_3 \text{ (\textit{solid})} + \text{heat} = CaO \text{ (\textit{solid})} + CO_2 \text{ (\textit{gas})}$$

2. *in common language:*

   quicklime + water = slaked lime + heat

   in scientific language:

   calcium oxide + water = calcium hydroxide + heat

   the formula:

   $$CaO \text{ (\textit{solid})} + H_2O \text{ (\textit{liquid})} = Ca(OH)_2 + \text{heat}$$

3. *in common language:*

   slaked lime + carbon dioxide = lime + water

   in scientific language:

   calcium hydroxide + carbon dioxide = calcium carbonate + water

   the formula:

   $$Ca(OH)_2 + CO_2 \text{ (\textit{gas})} = CaCO_3 \text{ (\textit{solid})} + H_2O \text{ (\textit{liquid})}$$

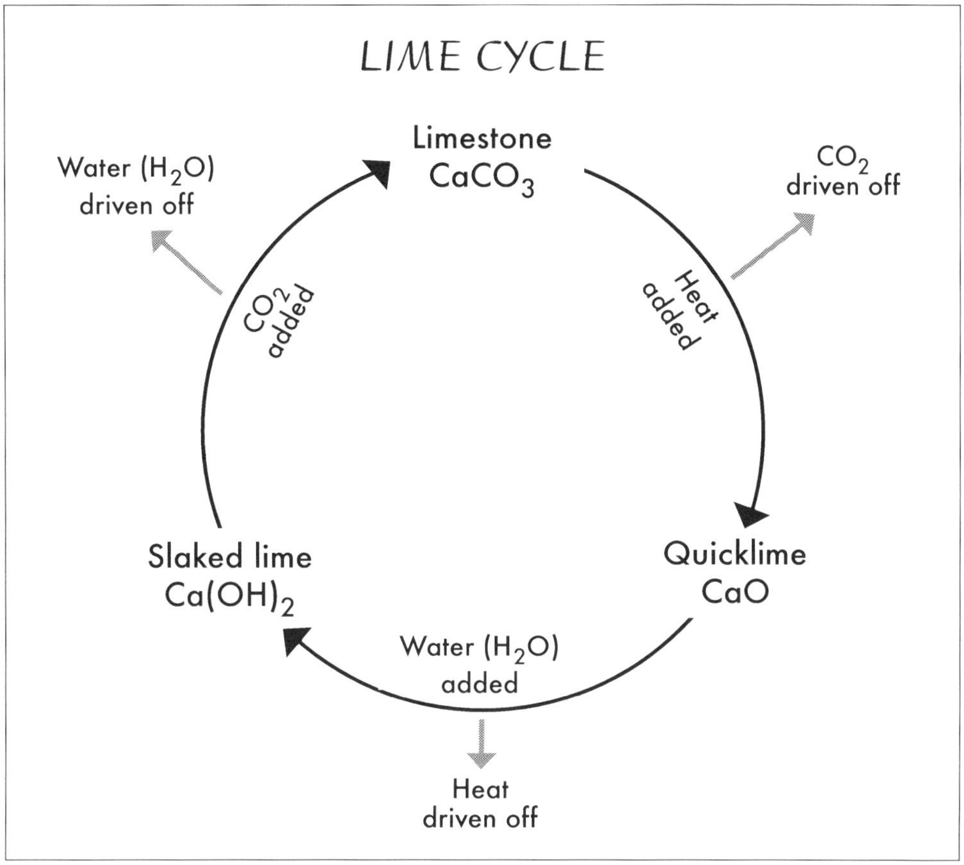

The burning of lime is an ancient process. Many cultures burned seashells, which contain lime, to create the quicklime needed for mortar. Others burned limestone, chalk, or marble. The Romans developed a particular type of mortar called pozzolana made of quicklime mixed with volcanic sand, which sets underwater, making it especially useful for building and repairing aqueducts. Lime kilns were originally simple structures that could be constructed at the building site, but as the process became industrialized, large permanent kilns were built.

There are several ways to heat lime with the students, and teachers will need to choose among them based on the resources and time available. If there is plenty of time and energy, the class can build a lime kiln. If the school has access to a farrier or blacksmith, lime can be burned in a forge. Alternatively, lime can be burned in the classroom with a propane or MAPP gas/oxygen torch.

Burning lime is interesting, but not very dramatic, because the lime's appearance only changes slightly. In contrast, slaking lime is a highly exothermic reaction, and students will be amazed by the changes in the shape of the lime and the prodigious release of steam as the lime absorbs water. The precipitation or hardening of lime takes time, but it is well worth making mortar to complete the cycle.

The study of lime establishes a foundation for the study of salt (sodium chloride) and for the study of acids and bases, which will be taken up next.

TOPICS

- The water cycle
- Changes of state
- Solid, liquid, gas
- Boiling and freezing points of water
- Electrical conductivity of water
- Electrical decomposition of water
- Burning, slaking, and setting lime

CHARACTERISTICS

- Water exists in its solid, liquid, and gaseous states in the earth's temperature range.
- Water goes through a cycle that includes precipitation, evaporation, and condensation.
- Water undergoes changes of state when heated and cooled.
- Water boils at 100° Celsius and freezes at 0° Celsius
- The boiling point of water can be raised by adding a solute.
- The freezing point of water can be lowered by adding a solute.
- Distilled water does not conduct electricity.
- Dissolving substances in water increases water's electrical conductivity.
- Water can be decomposed into hydrogen and oxygen.
- Limestone is composed of organic substances and exists in many forms.
- At high temperatures, limestone ($CaCO_3$) is converted into quicklime (CaO) with a release of carbon dioxide (CO_2).
- Adding water (H_2O) to quicklime (CaO) produces slaked lime ($Ca(OH)_2$).
- In the presence of carbon dioxide (CO_2), slaked lime turns into lime ($CaCO_3$).
- Mortar can be made using slaked lime, sand, and water.
- When limestone is in contact with a liquid acid, carbon dioxide is released.

DAY	DEMONSTRATIONS AND ACTIVITIES	CHARACTERISTICS
1	▪ The Water Cycle	▪ Water exists in its solid, liquid, and gaseous states in the earth's temperature range. ▪ Water goes through a cycle that includes precipitation, evaporation, and condensation.
	▪ Solid, Liquid, Gas	▪ Water undergoes changes of state when heated and cooled.
	▪ Determining the Boiling & Freezing Points of Water	▪ Water boils at 100° Celsius and freezes at 0° Celsius.
2	▪ Raising & Lowering the Boiling & Freezing Points of Water	▪ The boiling point of water can be raised by adding a solute. ▪ The freezing point of water can be lowered by adding a solute.
3	▪ Testing the Electrical Conductivity of Water & Solutions	▪ Distilled water does not conduct electricity. ▪ Dissolving substances in water increases water's electrical conductivity.
	▪ The Electrical Decomposition of Water	▪ Water can be decomposed into hydrogen and oxygen.
4	▪ The Lime Cycle	▪ Limestone is composed of organic substances and exists in many forms.
	▪ Burning Limestone	▪ At high temperatures, limestone ($CaCO_3$) is converted into quicklime (CaO) with a release of carbon dioxide (CO_2).
5	▪ Slaking Quicklime	▪ Adding water (H_2O) to quicklime (CaO) produces slaked lime ($Ca(OH)_2$).
	▪ Making Mortar	▪ In the presence of carbon dioxide (CO_2), slaked lime turns into lime ($CaCO_3$). ▪ Mortar can be made using slaked lime, sand, and water. ▪ When limestone is in contact with a liquid acid, carbon dioxide is released.

DAY 1

PRESENTATION

1. Water Cycle

- CHARACTERISTIC: Water exists in its solid, liquid, and gaseous states in the earth's temperature range.
- CHARACTERISTIC: Water goes through a cycle that includes precipitation, evaporation, and condensation.
- CHARACTERISTIC: Water undergoes changes of state when heated and cooled.

COMMENT

Water will be studied again in the eighth grade during the physics block, which considers water's physical properties and its role in weather and climate.

RELATED TOPICS

- Types of precipitation
- Drought
- Floods

DEMONSTRATION

2. Solid, Liquid, Gas

CHARACTERISTIC: Water exists in its solid, liquid, and gaseous states in the earth's temperature range.

In the sixth grade, this characteristic of water was introduced, but not investigated. Now students should learn the states of matter and the names of the processes that lead from one state to another.

PROCEDURE

- Place an Erlenmeyer flask on a hot plate or on a laboratory stand equipped with a ring clamp and a wire screen.
- Set up the Bunsen burner underneath the ring clamp.
- Place a few broken ice cubes inside the flask and put a whole one across the top of the flask.
- Heat the flask until the water begins to boil. Steam will rise to the ice cube on the top and some of the steam will condense in the neck of the flask.
- If necessary, put another ice cube on the top of the flask to create a kind of "rain" from the water vapor.

SAFETY

- Start off by heating the flask gently so that it doesn't crack.

COMMENT

Although this demonstration is very simple, its ramifications in the natural world are immense. These should be explored and discussed in depth during the review.

EQUIPMENT

Erlenmeyer flask, 500 ml
Bunsen burner
laboratory stand
ring clamp
wire screen

MATERIALS

ice cubes

LAB ACTIVITY

3. Determining the Boiling & Freezing Points of Water

CHARACTERISTIC: Water boils at 100° Celsius and freezes at 0° Celsius.

The boiling and freezing points of water are the basis for the Celsius thermometer. For these activities, students should record the Celsius temperatures as well as their Fahrenheit equivalents.

EQUIPMENT

test tubes
test tube rack
test tube clamp
thermometer
butane lighter
medium-sized beaker
stirring spoon
pliers

MATERIALS

chafing dish fuel
distilled water
crushed ice
salt

PROCESS

- **The Boiling Point of Water**
  - Pour about 1 inch of distilled water into a test tube.
  - Insert the thermometer so that the bulb is in the water and record the temperature.
  - Using a test tube clamp, hold the test tube at a slant of the flame so that the top of the water column begins to heat first.
  - Keeping the test tube in the flame, record the temperature at 1-minute intervals until the water is boiling.
  - Continue to leave the test tube in the flame for another minute and observe whether the temperature remains stable, rises, or falls.
- **The Freezing Point of Water**
  - Crush ice in a plastic bag with a hammer.
  - Create a slush of crushed ice and salt with about 5 parts ice to 1 part salt.
  - Fill the beaker a third full of slush.
  - Pour about 1 inch of distilled water into a test tube.
  - Insert the thermometer so that the bulb is in the water and record the temperature.
  - Put the test tube in the beaker with the slush.
  - Leaving the test tube in the beaker, record the temperature at 1-minute intervals until the water freezes.
  - Leave the thermometer in the test tube for another minute and observe whether the temperature remains stable, rises, or falls.

SAFETY

- If this is the first time that students have heated a test tube with liquid, they should be shown the proper way to heat it so that the water doesn't "burp" and spurt out of the test tube.
- Make sure that the test tube is always pointed away from anything and anyone else.
- Students should use pliers to replace the top of the chafing dish fuel.

COMMENT

Students should be told about the development of temperature scales and what led to the adoption of the Celsius and Kelvin scales.

RELATED TOPICS

- Types of thermometers
- Boiling and freezing points of other substances

DAY 2

LAB ACTIVITY

4. Raising & Lowering the Boiling & Freezing Points of Water

CHARACTERISTIC: The boiling point of water can be raised by adding a solute.

CHARACTERISTIC: The freezing point of water can be lowered by adding a solute.

EQUIPMENT
test tubes
test tube clamps
thermometer
beaker, medium
hammer or mallet
laboratory scoop
pliers

MATERIALS
chafing dish fuel
distilled water
salt
Ziploc plastic bag
ice
salt

PROCEDURE

- **Raising the Boiling Point of Water**
  - Holding the test tube at a slant, heat about an inch of distilled water in the test tube until it boils and record the temperature.
  - Remove the test tube from the flame, and using the scoop, add a small amount of salt to the water.
  - Continue to heat the test tube until the water boils again and record the temperature.
  - Add a little more salt to the solution and bring to the boiling point again.
  - See whether the boiling point can be raised even more.
- **Lowering the freezing point of water**
  - Crush ice in a plastic bag with a hammer.
  - Create a slush of ice and salt with about 5 parts ice to 1 part salt.
  - Fill the beaker one third full of slush.
  - Pour about 1 inch of distilled water into a test tube.
  - Insert the thermometer so that the bulb is in the water and record the temperature.
  - Put the test tube in the beaker with the slush.

- Record the temperature as the water starts to freeze around the edges.
- Using the scoop, add a pea-sized portion of salt to the water and stir with the thermometer.
- See if the water is still liquid at 0° Celsius
- Add another pea-sized portion of salt to the water and stir.
- Allow the water to freeze, and record the temperature at which it freezes. (The water may not freeze but will be cooled to at least –10°C.)

SAFETY

- If you are crushing the ice yourself, make sure that the bag is well sealed before hammering.
- Students should use pliers to replace the top of the chafing dish fuel.

RELATED TOPICS

- Commercial and industrial applications of this principle
- Use of salt for melting ice and snow

DAY 3

DEMONSTRATION

5. Testing the Electrical Conductivity of Water & Solutions

CHARACTERISTIC: Distilled water does not conduct electricity.

CHARACTERISTIC: Dissolving substances in water increases water's electrical conductivity.

This demonstration can also be done as a lab activity.

EQUIPMENT
pliers
wire strippers (optional)
beakers, medium
dropper or plastic pipette

MATERIALS
6-volt battery
insulated bell wire, 4′ long
small 1.5-volt bulb
bulb socket
distilled water
salt
hydrochloric acid
sodium hydroxide (lye)

PROCEDURE

- **Setup**
  - Cut the wire into three pieces, two 12-inch pieces and one 24-inch piece.
  - Strip the ends of the wires. The wires are now called leads.
  - Connect the 12-inch leads to the bulb base and connect one of them to the battery.
  - Connect the 24-inch lead to the other terminal of the battery.
  - Test the circuit by touching the ends of the loose wires together.
  - If the bulb lights up, then you are ready to test for the conductivity of water.
- **Testing the conductivity of distilled water**
  - Fill the three beakers three-quarters full of water.
  - Put the unattached ends of the leads into the water, without the ends touching each other.
  - If the bulb lights up, the water is conducting electricity and the beaker needs to be rinsed thoroughly and filled with distilled water.
  - If the bulb does not light up, try bringing the leads closer together.

  - The bulb should not light up until the leads are very close or touching.
- **Testing the conductivity of various substances**
  - Fill the beakers three-quarters full of water.
  - Dissolve about ½ teaspoon of salt in one, about ½ teaspoon of sodium hydroxide in another, and about 10 drops of hydrochloric acid in a third.
  - Test the solution in each beaker using the leads and bulb to reveal whether the solution conducts electricity.

SAFETY

- Everyone working with sodium hydroxide or hydrochloric acid must wear goggles, gloves, and aprons.
- The teacher should mix and pour the acid and base solutions to be tested.
- When pouring a solution from a flask into a beaker, make sure that the beaker is placed on a stable surface. Tilt the flask so that its lip rests on the lip of the beaker and then continue to tilt until the proper amount of liquid has been transferred.

COMMENT

Electrolytes have myriad uses and applications. Students could research some of these and report on what they have learned.

DEMONSTRATION

6. The Electrical Decomposition of Water

CHARACTERISTIC: Water can be decomposed into hydrogen and oxygen

EQUIPMENT

electrolysis apparatus

MATERIALS

water
salt or other electrolyte

PROCEDURE

- Refer to the instructions included with the electrolysis apparatus to learn how to use it to decompose water.
- Have students observe the rate of bubbling around the anode and cathode and the volumes of gas collected.
- If sufficient hydrogen and oxygen have been collected, the presence of oxygen and hydrogen can be revealed by combustion.

SAFETY

- Teachers should wear goggles, gloves, and an apron when handling acidic or alkaline electrolytes.
- The electrolyte solution can be diluted and safely flushed down the sink.

COMMENT

This demonstration provides a good introduction to chemical formulas, because the ratio of hydrogen to oxygen is so clear.

DAY 4

PRESENTATION

7. Limestone: Its Sources & Uses

CHARACTERISTIC: Limestone is composed of organic substances and exists in many forms.

The students should have learned about limestone in their sixth-grade study of geology. Review some of the different types of limestone (travertine, tufa, and chalk, as well as oolitic, fossiliferous, and lithographic limestone) and describe how they are formed. Review the process by which limestone is metamorphosed into marble.

Students should be told how mollusks create their shells out of calcium carbonate and how, for millennia, people used seashells to create the lime needed for mortar.

Some description of the history of mortar may also be of interest.

ACTIVITY

8. Burning Limestone

CHARACTERISTIC: At high temperatures, limestone ($CaCO_3$) is converted into quicklime (CaO) with a release of carbon dioxide (CO_2).

PROCEDURE

- Begin by creating quicklime. To create the best quicklime, use small pieces of the purest limestone or marble you can find. The limestone must be brought up to 900°C (1,650°F) to go through the chemical change.
- Here are three possibilities for heating limestone.

  - **Kiln:** A lime kiln allows you to burn larger quantities of lime. The class can build a lime kiln out of fire brick. This is an exciting and rewarding project, but it requires a significant time commitment. Plans for a lime kiln are provided by John Hofman in David Mitchell's *The Wonders of Waldorf Chemistry*, available for download from www.waldorflibrary.org.
  - **Forge:** A blacksmith's or farrier's portable or permanent forge can be used to burn small pieces of limestone.
  - **Classroom:** A MAPP gas/oxygen torch can be used to burn very small pieces of limestone.
- Use the method that is most practical for you. You will need to produce at least a few small pieces of quicklime to use in the following activities.

DAY 5

ACTIVITY

9. Slaking QuickLime

CHARACTERISTIC: Adding water (H_2O) to quicklime (CaO) produces slaked lime ($Ca(OH)_2$).

EQUIPMENT
baking pan or glass baking dish
watch glass
flask or jar
plastic pipette or dropper

MATERIALS
burned lime (quicklime, made in class)
water

If you have made a large quantity of quicklime, then this can be done on a larger scale. If you only have small pieces, they can be placed on a watch glass.

PROCEDURE

- Put pieces of quicklime on a baking sheet or in a glass baking dish.
- Slowly pour some water onto the lime.
- Observe the reaction of the lime as it reacts to water. It should absorb the water and then emit steam.
- Add a little more water, but do not add so much that the lime becomes a slurry.

- If this is done with a small piece of quicklime on a watch glass, students can feel the heat of the reaction through the bottom of the glass.

SAFETY

- Students need to wear goggles and gloves for this activity.
- The slaking process is very exothermic. Students will be able to feel the heat by putting their hands above (but not touching) the slaking lime, but they should be ready to lift their hands if the heat is excessive.

ACTIVITY

10. Making Mortar

CHARACTERISTIC: In the presence of carbon dioxide (CO_2), slaked lime turns into lime ($CaCO_3$).

CHARACTERISTIC: Mortar can be made using slaked lime, sand, and water.

PROCEDURE

- Grind the slaked lime into a powder
- Mix slaked lime and fine sand in the proportion of 1: 3.
- Add enough water to make a putty-like mortar.
- Use the mortar to lay some bricks or to cover a surface.
- In order to harden, the mortar will need to be exposed to the carbon dioxide in the air, which can take several days.

SAFETY

- Students should wear goggles and gloves for this activity.

EQUIPMENT

mortar and pestle
bowl or small tub
stirring spoon

MATERIALS

slaked lime
fine sand
water

DEMONSTRATION

11. Revealing the Carbon Dioxide Contained in Limestone

CHARACTERISTIC: When limestone is in contact with a liquid acid, carbon dioxide is released.

Students should be told that the gas that is bubbling out is carbon dioxide.

If the reaction is allowed to continue for a while, one could test the gas above the solution by lowering a match into the flask. The match should be extinguished.

EQUIPMENT

Erlenmeyer flask, small

MATERIALS

dilute hydrochloric acid

small pieces of limestone

PROCEDURE

- Fill a small flask with slightly diluted hydrochloric acid. (See the note about safety and be careful to follow the correct order.)
- Drop a small piece of limestone in the flask and observe that gas bubbles are released from the limestone.

SAFETY

- The teacher should wear goggles and gloves for this demonstration
- To dilute the acid, pour the acid into the water, not the other way around.
- The hydrochloric acid can be diluted, neutralized with baking soda, and safely flushed down the sink.

COMMENT

During the review, the three stages of the lime cycle should be reviewed.

Week 4 of Grade 7 Chemistry

SALTS, ACIDS, & BASES

During the final week of our block, we will examine sodium chloride, a common salt that provides us with essential nutrients. It also contains the raw material for hydrochloric acid and for sodium hydroxide, which are among the most useful of all acids and bases. Be sure your students understand the importance of salt in daily life, and even to life itself.

Salt is an essential nutrient that helps us preserve the fluid balance in our cells. Our kidneys help maintain a healthy salt balance, expelling excess salt through our urine. We also get rid of salt by sweating.

Salt is used as a seasoning because it brings out the flavor in foods, but before the advent of refrigeration, its major use was for food preservation. Salt was one of the first articles to be traded and taxed, and it served as a status symbol, separating the nobility from those who sat "below the salt" at the banquet table.

One of the most common uses for salt today is for melting snow and ice, and about 20 million tons a year are used for that purpose. Because salt lowers water's freezing point, the briny water melts the ice and stops new ice from forming.

Before working with salt in the lab, it's important that the students learn about how salt is formed, how it is obtained, how it has been used throughout history. Salt comes from two main sources: halite and seawater. Halite, or rock salt, was formed when enclosed oceans, lakes, and playas evaporated, leaving behind salt sediments, which can be hundreds of feet thick. The Great Salt Lake in Utah and the Dead Sea between Jordan and Israel are two of many hypersaline bodies of water where halite is still being formed today.

The salt in the ocean mostly comes from runoff that contains dissolved ions from the many minerals that contain sodium and chlorine. Hydrothermal vents and submarine volcanism also contribute dissolved ions to sea water. To give students a sense of how much dissolved salt is contained in the oceans, they can be told that if all the salt in the ocean were spread over the earth's land surface, it would form a layer over 500 feet thick.

Throughout the ages, people have obtained salt by harvesting and mining. The most common method of harvesting salt, which is still practiced, uses a series of shallow basins that are exposed to the sun and wind. As the water evaporates, the salt concentration increases until the salt precipitates and can be gathered.

Rock salt can be mined like any other mineral. The mining used to be done by hand but now is mostly done with machines. Salt can also be extracted by drilling shafts into the salt deposit, pumping hot water into the shafts, extracting the saline solution, and evaporating it in vacuum pans.

Salt—sodium chloride—is composed of sodium and chlorine. Sodium— isolated by Humphry Davy in 1807—is a soft, silvery metal that bursts into flame when it comes in contact with water. Chlorine—isolated by Carl Wilhelm Scheele in 1774—is a poisonous greenish gas. Sodium is the sixth most common element on earth and chlorine the twentieth, yet both of these elements are never found in nature in their elemental form because they are so reactive. It's hard to believe that salt, such an inert, benign substance, is made of such reactive elements!

In the past two hundred years, we have found innumerable uses for chlorine and sodium. Chlorine was first used as a bleaching agent for fabrics and paper, as well as a disinfectant, but now it is also used in the production of vinyl, solvents, herbicides, and pesticides. Hydrochloric acid, one of chlorine's most common compounds, is used in the production of petroleum, steel, industrial cleaning products, and many pharmaceuticals.

Sodium also has myriad uses. Liquid sodium serves as a coolant in nuclear reactors. Compounds of sodium are used in the production of aluminum, glass, and paper and to transform cellulose into rayon. Sodium compounds are also used in soaps, detergents, and shampoos, as antimicrobial agents and as medicines to treat acid indigestion and strengthen teeth.

Having given the students a wide-ranging introduction, we can examine salt's physical properties, and we can try to burn it. A propane torch is not hot enough, but we can use a MAPP gas/oxygen torch to melt salt. The golden-yellow flame that it produces is sodium's elemental "signature."

Salt is highly soluble, and students will learn how to create saturated and supersaturated solutions. Then they will observe the processes of precipitation and crystallization, which are delicate and beautiful, quite the opposite of the dramatic combustion reactions of the first two weeks.

Our work with combustion provides the basis for the introduction of acids and bases. When the gases that are produced by combustion combine with water, they produce an acid, while the ash that remains is alkaline. The word "alkali" comes from the Arabic *al-qaliy*, which refers to the burnt ashes of a plant.

Although students don't need to know many chemical formulas at this age, it would be good for them to learn the chemical formulas for salt ($NaCl$), hydrochloric acid (HCl), and sodium hydroxide ($NaOH$). Having observed how hydrochloric acid and sodium hydroxide creates salt ($NaOH + HCl = NaCl + H_20$), they can be told that an aqueous solution of salt ($NaCl + H_20$) can be separated into hydrochloric acid (HCl) and sodium hydroxide ($NaOH$) plus water. This is also a good time to mention, and perhaps show them, the periodic table of the elements. It is important that students learn that most elements are found in compounds rather than in their elemental forms.

By feeling, smelling, and tasting very dilute acidic and basic solutions, students will recognize that these substances are opposites in many ways. Acids are volatile, bases stable. Acids have a sharp smell and taste (the word "acid" derives from the Latin word *acere*, which means to be sour or sharp), while bases taste insipid; acids feel grainy, bases slippery.

Students will also observe that acids and bases have contrasting effects on various substances: acids "cook" or "burn" the substance, while bases digest them. This is a good opportunity to remind the students that our stomachs contain hydrochloric acid of similar strength to the concentrated acid we use in the lab, and our gall bladder, pancreas, and intestines produce alkali digestive juices.

Indicators reveal the presence of acids and bases by changing color, and students will be delighted to work with red cabbage juice, a readily made, very sensitive indicator.

Electricity can be generated by the interaction of acids and metals. If the generation of electricity was not covered in the physics block, generating electricity can serve as our final lab activity in chemistry. Students can be told about the discovery of current electricity by Luigi

Galvani and Alessandro Volta, and then they can generate electricity themselves with copper, zinc, and an acidic solution. At this age, it is not important for students to understand how electricity arises through the interaction of the metals and the electrolyte, but their hands-on experiences will provide a firm foundation for the explanations they will be taught in high school and beyond.

I recommend ending the chemistry block with a beautiful demonstration that will help balance the students' increasingly materialistic view of the physical world. Just like sodium with its distinctive color, each element has a signature color when it is heated, and students will be amazed and delighted by the final demonstration of the colored flames produced by various salts.

Ideally, the seventh-grade chemistry block will satisfy the students' hunger to learn more about the world and its substances, and it should whet the students' appetite for further work with this interesting and captivating subject in eighth grade.

TOPICS

- Salts
- Solutions
- Acids and bases
- Generating electricity

CHARACTERISTICS

- Salt has distinctive physical properties.
- Salt has a high melting point and burns with a golden flame.
- Salt can be dissolved in water.
- A solution of salt in water can be saturated or supersaturated.
- Salt can be precipitated out of a solution by cooling or by the addition of a compound containing one of the elements in salt.
- Acids and bases have distinctive physical and chemical properties.
- Acids and bases react with many organic substances.
- Indicator solutions reveal the presence of an acid or a base.
- Electricity can be generated when two different metals are in contact in the presence of an electrolyte.
- The presence of electricity can be perceived by our sense of taste.
- A galvanometer can be used to reveal the presence of current electricity.

- A voltaic cell generates an electrical current.
- Each element combusts with a characteristic color.

DAY	DEMONSTRATIONS AND ACTIVITIES	CHARACTERISTICS
1	■ Salt (Presentation) ■ Physical Properties of Salt	■ Salt has distinctive physical properties. ■ Salt has a high melting point and burns with a golden flame.
2	■ Dissolution & Precipitation of Salt	■ Salt can be dissolved in water. ■ A solution of salt in water can be saturated or supersaturated. ■ Salt can be precipitated out of a solution by cooling or by the addition of a compound containing one of the elements in salt.
3	■ Properties of Acids & Bases ■ The Effects of Acids & Bases on Organic Substances	■ Acids and bases have distinctive physical and chemical properties. ■ Acids and bases react with many organic substances.
4	■ Identifying Acids & Bases with an Indicator Solution	■ Indicator solutions reveal the presence of an acid or a base.
5	■ Detecting Electricity through Taste ■ Detecting Electricity with a Galvanometer ■ The Voltaic Cell	■ Electricity can be generated when two different metals are in contact in the presence of an electrolyte. ■ The presence of electricity can be perceived by our sense of taste. ■ A galvanometer can be used to reveal the presence of current electricity. ■ A voltaic cell generates an electrical current.
6	■ Burning Salts	■ Each element combusts with a characteristic color.

DAY 1

PRESENTATION

1. Salts

CHARACTERISTIC: Salt exists in many forms, is extracted through many processes, and has many uses.

DEMONSTRATION

2. Physical Properties of Salt

CHARACTERISTIC: Salt has distinctive physical properties.

CHARACTERISTIC: Salt has a high melting point and burns with a golden flame.

EQUIPMENT
microscope
deflagration spoon
Bunsen burner
lab striker or butane lighter
propane torch or MAPP gas/oxygen torch
lab stand
utility clamp
laboratory safety shield

MATERIALS
several types of salt
dark paper

PROCEDURE

- Show students the physical properties of salt: its appearance, texture, odor, and flavor. If salt crystals are observed on dark paper, they can be seen better.
- If possible, have students examine different types of salts under a microscope or look at enlarged photographs of salt crystals.
- Mount the deflagration spoon on the lab stand with the utility clamp approximately five inches above the Bunsen burner cylinder.
- Heat salt in the deflagration spoon from below with a Bunsen burner and from above with a propane or MAPP gas/oxygen torch. The salt will melt and burn. The bright yellow-orange flame is a characteristic of the metal sodium, which is one of the components of sodium chloride.

SAFETY

- The teacher should wear goggles, gloves, and an apron and should do the demonstration behind the laboratory safety shield.

COMMENT

A soup spoon can be used as a deflagration spoon.

Minor deposits on the spoon left by heating salt can be cleaned with sandpaper and steel wool.

Major deposits can be cleaned by heating the spoon until it is red hot and then quenching it in cold water.

DAY 2

DEMONSTRATION

3. Dissolution & Precipitation of Salt

CHARACTERISTIC: Salt can be dissolved in water.

CHARACTERISTIC: A solution of salt in water can be saturated or supersaturated.

CHARACTERISTIC: Salt can be precipitated out of a solution by cooling or by the addition of a compound containing one of the elements in salt.

Prior to this demonstration, prepare some supersaturated salt solution for use later in this demonstration.

PROCEDURE

- Put 50 ml of cold distilled water into a 250-ml Erlenmeyer flask. Pour in a small amount of salt and stir, then continue to add more salt until no more dissolves. When the water cannot absorb any more salt, the solution is considered to be saturated.

EQUIPMENT

1 Erlenmeyer flask, 250 ml
2 Erlenmeyer flasks, 125 ml
laboratory stand
ring clamp with screen
utility clamp
stirring rod
4 test tubes
test tube rack
beaker, medium
dropper or pipette
laboratory safety shield
funnel

MATERIALS

salt
cold water
dark paper
hydrochloric acid
sodium hydroxide pellets
filter paper

- **Setup**
  - Attach a ring clamp to a laboratory stand and place a wire screen on the clamp.
  - Place the flask on the screen and secure the neck of the flask with a utility clamp.
  - Set up the Bunsen burner beneath the flask.
- **Process**

  Precipitating salt out of solution by cooling
  - Heat the saturated solution until all the salt that is on the bottom dissolves. Add more salt and stir until no more dissolves.
  - The hot solution that cannot absorb more salt is considered to be supersaturated.
  - Remove the flask with the supersaturated solution from the laboratory stand and pour some of the supersaturated solution into two 125-ml Erlenmeyer flasks. Fill several test tubes with about 2 inches of supersaturated salt solution. Place the test tubes in a test tube rack.
  - Cool one of the test tubes by placing it in a beaker of cold water.
  - Have students observe the precipitation of salt out of the solution.
  - Leave the other test tubes and flasks to cool naturally; salt crystals should begin to grow on the bottom and up the sides.

  Precipitating salt out of solution by adding acid
  - Fill a 125- or 250-ml Erlenmeyer flask about three-quarters full of the cooled, supersaturated salt solution.
  - Set up a dark background behind the flask, dim the lights, and illuminate the flask from the side.
  - Drop hydrochloric acid into the solution and have the students observe the snowflake-like precipitation of salt from the solution.

  Combining hydrochloric acid and sodium hydroxide
  - Pour about ¾ inch of concentrated hydrochloric acid into a test tube.
  - Put in a pellet of sodium hydroxide and observe the reaction. Add additional pellets.
  - Allow the test tube to cool and sit overnight. The next day, strain the liquid through a funnel with filter paper and remove the salt that has been created.

SAFETY

- Goggles, gloves, and an apron should be worn to perform these demonstrations.
- The hydrochloric acid should be poured into a small beaker and covered with a watch glass.
- When pouring the acid into the test tube, make sure that the test tube is stable. Then place the lip of the beaker against the rim of the test tube and slowly tilt the beaker to pour out the acid.
- The reaction of concentrated hydrochloric acid with pellets of sodium hydroxide is vigorous and exothermic. Use a laboratory safety shield.
- Do not taste the salt created by the hydrochloric acid and sodium hydroxide because it may have dangerous residues.
- The acid-base mixture can be diluted and be safely flushed down the sink.

DAY 3

Acids and bases are byproducts of combustion, with the smoke being acidic and the ash being alkali or basic. If there is time, demonstrate this to the students using moist indicator paper or assign it to a student to do as an independent project.

ACTIVITIES

4. Properties of Acids & Bases

CHARACTERISTIC: Acids and bases have distinctive physical and chemical properties.

PROCEDURE

- Create the following solutions in the beakers and cover each beaker with a watch glass.
  - dilute solution of hydrochloric acid

EQUIPMENT
4 medium-sized beakers
watch glasses
droppers or pipettes

MATERIALS
hydrochloric acid
sodium hydroxide
water

  - highly diluted solutions of hydrochloric acid (See note on taste below.)
  - dilute solution of sodium hydroxide
  - highly diluted solution sodium hydroxide. (See note on taste below.)
- **Texture**
  - Have students dip their forefinger in the dilute hydrochloric acid solution and then rub the liquid between their forefinger and thumb. The dilute acid will have a dry, gritty texture. They should rinse and dry their hands afterwards.
  - Repeat with the solutions of sodium hydroxide, which is a base. Bases have a slippery, soapy texture. When they are done, students should again rinse and dry their hands.
- **Odor**
  - Securely hold a beaker of dilute hydrochloric acid in front of each student. Have the student wave her hand to gently waft the air above the vapors towards her face. If the vapor is too intense, she can simply move back. The acid will have an acrid smell.
  - Repeat with the base. If students can smell it at all, it will have a bland, soap-like smell.
- **Taste**
  - For this, you need to use the highly dilute solutions and to test them yourself to make sure they are the right strength. The taste should be perceptible but not irritate the tongue.
  - Have the students dip the tip of a finger in the solution and put a tiny drop on their tongue. If they can't taste anything, have them do it again.
  - The acid will taste sour, the base soapy.

SAFETY

- Teachers and students must wear goggles during this activity.
- Always put the acid or the base into the water. Do not add water to the acid or the base.
- Test the solutions for the proper level of dilution before having the students smell, touch, or taste them.
- To experience the smell, hold the beakers with both hands so that they are stable.

- One rarely tastes anything in a chemistry lab; this experience is an exception to give the students an opportunity to learn an essential quality of the substances.
- For the touch and texture experiences, hold the beakers firmly on a table or desk.
- To be disposed of, the acid and base can be diluted and safely flushed down the sink.

RELATED TOPICS

- Different types of acids and bases and their properties
- Use of acids in foods and beverages
- Use of bases in household products

DEMONSTRATION

5. The Effects of Acids & Bases on Organic Substances

CHARACTERISTIC: Acids and bases react with many organic substances.

PROCEDURE

- **Setup**
  - Pour a small amount (5 to 10 ml) of concentrated sulfuric acid into a beaker and cover it with a watch glass.
  - Prepare a small amount (5 to 10 ml) of very concentrated sodium hydroxide in a beaker and cover it with a watch glass.
  - Pour a small amount of concentrated hydrochloric acid into a beaker and cover it with a watch glass.
  - Set up two rows of portion cups in a tray, with about 10 to 12 in each row and identify one row as acid and the other as base.

EQUIPMENT

small beakers
watch glasses
2 medium beakers
3 watch glasses
laboratory safety shield

MATERIALS

concentrated sulfuric acid
sodium hydroxide
concentrated hydrochloric acid
water
piece of paper
wooden splint
piece of cloth from natural material
24 plastic small-portion cups with lids
variety of food substances, including carbohydrates, fats, and proteins
plastic wrap

- **Process**

  Effect of sulfuric acid on organic substances

  - Using a pipette, place a few drops of the concentrated sulfuric acid on a piece of paper that is on a watch glass.
  - Place a few drops on a wooden splint.
  - Place a few drops on a piece of cotton, silk, linen, or wool fabric.
  - When you are done with the sulfuric acid, remove it from the demonstration area.

  Effect of hydrochloric acid and sodium hydroxide on organic substances

  - Put a small amount of each of the food substance to be tested in the shot glasses in the acid row and the shot glasses in the base row.
  - Add enough concentrated hydrochloric acid to cover the substances in the acid row.
  - Add enough concentrated sodium hydroxide to cover the substances in the base row.
  - Cover the rows of shot glasses with plastic wrap and put in a place where they will not be disturbed.
  - During the course of the day, students may examine what is happening to the substances, but the closer examination will happen the next morning.

SAFETY

- Everyone should wear goggles and gloves for these demonstrations, and the teacher should wear an apron.
- Students should observe the demonstrations with sulfuric acid from behind the laboratory safety shield.
- To dispose of any remaining sulfuric acid, it should be diluted by pouring it into water, neutralized with baking soda or dilute sodium hydroxide, and flushed down the sink.
- Do not pour water into the beaker with the sulfuric acid; pour the acid into the water.
- To dispose of any remaining hydrochloric acid, it should be diluted by pouring it into water, neutralized with baking soda, and flushed down the sink.
- To dispose of any remaining sodium hydroxide, it should be diluted and flushed down the sink.

- The pipettes can be safely disposed of in the trash.

RELATED TOPICS

Industrial and commercial uses of acids and bases

DAY 4

DEMONSTRATION

6. Identifying Acids & Bases with an Indicator Solution

CHARACTERISTIC: Indicator solutions reveal the presence of an acid or a base.

PROCEDURE

- **Setup**
  - Prepare red cabbage indicator solution in advance by chopping up red cabbage leaves, covering them with distilled water, and boiling the mixture for about 10 minutes.
  - The solution should be a rich bluish-purple color. It can be diluted to make it lighter in hue and more sensitive.
  - Prepare a concentrated solution of sodium hydroxide in a small beaker and cover it with a watch glass.
  - Pour hydrochloric acid into a small beaker and cover it with a watch glass.
- **Process**
  - Fill two large flasks or cylinders with red cabbage solution.
  - Using a pipette, slowly drop concentrated hydrochloric acid into one of them until the solution begins to turn red.
  - Do not stir. Let the color change proceed gradually.
  - Slowly drop sodium hydroxide in the other flask or cylinder.
  - Other foods and liquids can be tested to determine whether they are acidic or basic.

EQUIPMENT

large pot
stove
large flasks or cylinders
2 medium beakers
2 watch glasses
2 droppers or pipettes

MATERIALS

red cabbage
water
concentrated hydrochloric acid
sodium hydroxide
cabbage juice indicator solution
clear plastic cups
various foods and liquids

- The speed and degree of color change will also indicate whether they are a strong acid or a base.

SAFETY

- Wear goggles and gloves to perform these demonstrations.
- Any remaining hydrochloric acid should be diluted, neutralized with baking soda, and flushed down the sink.
- Any remaining sodium hydroxide should be diluted and flushed down the sink.
- The used indicator solution should be diluted and flushed down the sink.
- Refrigerate the leftover cabbage juice because it spoils quickly.

COMMENT

Red cabbage indicator solution is the easiest to prepare and to work with, but a universal indicator solution available from a biological or chemical supply company can be used.

Students can be given a leaf of red cabbage to take home to create their own indicator solution, which they can use to test common foods and household products.

In preparation for the review the next day, it's lovely to prepare a set of seven clear plastic cups with cabbage juice that has been tinted with different strengths of acids and bases to create a spectrum. Arrange the cups on a windowsill so that the natural light highlights the colors.

RELATED TOPICS

- Other types of fruit and vegetable indicators
- Use of indicators, e.g., to check the pH of a swimming pool

DAY 5

ACTIVITY

7. Detecting Electricity Through Taste

CHARACTERISTIC: Electricity can be generated when two different metals are in contact in the presence of an electrolyte.

CHARACTERISTIC: The presence of electricity can be perceived by our sense of taste.

Introduce these activities by a brief presentation about the discoveries of Luigi Galvani and Alessandro Volta.

PROCEDURE

- Prepare the copper and zinc strips by polishing them with steel wool.
- Have students "taste" electricity by holding a piece of zinc and a piece of copper on the sides of their tongues and allowing the other ends of the metal pieces to touch each other.
- The sour taste that they experience is evidence of the flow of electricity.

SAFETY

- If you don't have enough copper and zinc strips for each student, wash the strips thoroughly between uses.

MATERIALS

strips of copper, 1″x 4″

strips of zinc, 1″ x 4″

steel wool

LAB ACTIVITY

8. Detecting Electricity with a Galvanometer

CHARACTERISTIC: A galvanometer can be used to reveal the presence of current electricity.

MATERIALS

map compasses

about 4′ of magnet wire, 22- or 24-gauge

D or C cell batteries

PROCEDURE

- Leaving 18 inches of magnet wire free on one end, wind the wire across the width of the map compass. Keep the coils tight and neat.
- Stop coiling when there are 18 inches of magnet wire left at the other end.
- It is helpful to hook that length of magnet wire once or twice through the other coils so that they don't become loose.
- Strip 1 inch of the enamel coating from the ends of the magnet wire by holding them in the flame of a match for a few seconds.
- To use the galvanometer, hold the compass flat and move it so that the needle is parallel to the coils of wire.
- When the ends of the wire are connected to the poles of the battery, the compass needle will deflect from its north-south orientation.

LAB ACTIVITY

9. The Voltaic Cell

CHARACTERISTIC: A voltaic cell generates an electrical current.

PROCEDURE

- **Setup**
  - Prepare the electrolyte solution in advance by mixing 1 part of

concentrated sulfuric acid into 20 parts water in a flask.

- Add a very small pinch (less than 1/8 teaspoon) of potassium dichromate to the solution.
- Stopper the flask.
- *Note: The potassium dichromate is a depolarizer that stops hydrogen bubbles from clinging to the zinc electrode that will be inserted. If you decide not to use the compound because it requires special disposal, be prepared to periodically brush the hydrogen bubbles off the zinc electrode with a stirring rod or splint as you proceed.*
- Set up a fan in the window to improve air circulation
- Check that the copper and zinc strips are not tarnished. If they are, have the students polish them with steel wool.
- Cut slits on opposite sides of the coffee cup covers so that the metal strips can be inserted through the slits and hang inside the beaker. You will insert a zinc strip and a copper strip through each coffee cup cover.
- Bend the top inch of the metal strips so that they can be draped over the sides of the beakers while the longer parts of the strips hang down into the beaker.
- Insert the strips through the coffee cup covers and attach the alligator clips to the ends of the metal strips.
- Do not place the strips in the beaker until the solution has been poured.

- Process
  - The teacher should pour the dilute acid solution into the beaker so that it fills the beaker three-quarters full. Cover the beaker with the coffee cup cover and insert the metal strips so that they are against the side of the beaker and do not touch each other. At this point, the voltaic cell is complete. Note that you may need four to six voltaic cells to create the "crown of cups" that is described below.
  - The remaining processes should be done by the students.
  - Hook up the voltaic cell to the galvanometer to see if the compass needle is deflected.
  - If it is not, and the galvanometer worked when you tested it with the battery, check the connections of the leads to the metal strips.

EQUIPMENT

Erlenmeyer flask, 500 or 1000 ml
medium beakers for each group (8-ounce jelly jars can be substituted)
plastic coffee cup covers
wire leads with alligator clips, 2 per group
low-voltage electric motors
buzzers
1.5-volt small light bulbs

MATERIALS

sulfuric acid
potassium dichromate
copper and zinc strips
steel wool

- Hook up the wire leads to the low-voltage motor.
- To make the motion of the motor more apparent, attach a small paper propeller to the motor shaft.
- If time allows, create a crown of cups as follows:
  - Arrange the voltaic cells in a circle so that the wires can reach from one voltaic cell to the next.
  - Using the wire leads with the alligator clips, attach the wire from the zinc strip of the first voltaic cell to the copper strip of the second voltaic cell. The wire from the zinc strip of the second cell goes to the copper strip of the third cell, etc.
  - By the end you should have two free wires: one from the copper of the first cell and one from the zinc of the last cell. These two wires can be hooked up to the motor, buzzer, or a small low-voltage bulb.

COMMENT

If you have a voltmeter, you can check the voltage of the individual cells and the crown of cups.

When the cell is not being used, remove the metal strips; otherwise the zinc will continue to be dissolved by the acid.

RELATED TOPICS

- Other types of wet cells
- Use of wet cells in early telegraphy
- Dry cells

SAFETY

- Everyone who is working with the voltaic cells should wear goggles and latex gloves.
- If the acid solution gets on someone's skin, it should be rinsed off with plenty of water.
- Potassium dichromate needs to be disposed of as a hazardous waste. Check local ordinances regarding safe disposal.
- If you have used a sulfuric acid solution, rather than one that contains potassium dichromate, it can be disposed of by further diluting with water, neutralizing with baking soda or dilute sodium hydroxide, and safely flushing it down the sink.

DAY 6

DEMONSTRATION

10. Colored Flames of Salts

CHARACTERISTIC: Each element combusts with a characteristic color.

PROCEDURE

- **Setup**
  - Prepare dilute solutions of the salts and pour them into the spray bottles. Label the bottles.
- **Process**
  - Light the Bunsen burner and adjust the flame so that it is tall.
  - Spray a little sodium chloride solution into the flame.
  - The students may recognize the color from their experience of burning salt.
  - Repeat with the other salts.

SAFETY

- Everyone should wear goggles, gloves, and an apron for this demonstration.
- You may decide to do this behind the laboratory safety shield, but it is not necessary if the students stand far enough away from the flame.
- The chloride solutions are not dangerous but should be rinsed off it they get on anyone's skin.
- The chloride solutions can be stored in spray bottles for future use, or they can be diluted and safely flushed down the sink.

COMMENT

This demonstration can be done during the review on the Monday after the block is over.

EQUIPMENT

small spray bottles

Bunsen burner

lab striker or butane lighter

MATERIALS

small amounts of the following salts:

sodium chloride

barium chloride

calcium chloride

copper chloride

lithium chloride

potassium chloride

strontium chloride

Save the most dramatic colors for last. End with strontium or lithium (magenta) and copper (blue/green).

Students may ask for more than one salt to be sprayed into the flame. I recommend doing a few combinations, but do not overdo it, because the impressions become duller.

RELATED TOPICS

- Spectral analysis
- Salts used in fireworks, flares, etc.

Chapter 10: Eighth Grade Chemistry

How does a wheel—with its hub, spokes, and rim—serve as a metaphor for our work with the students in the eighth grade? The teacher is the hub of the wheel. She holds the vision for the class, she determines the content, and she chooses the methods that will best serve the students.

The community—and, by extension, the world—is the rim. What happens in the classroom should eventually have an impact on the community and the world. Conversely, what happens outside of school should find its way into the classroom.

The students are the spokes, mediating between the hub and the rim. Through their learning and working, students should carry the impulses from their teachers into the community and the world. Through their encounters and relationships, students should bring their experiences of the world back to their class and their teachers.

The wheel is a unified whole, with each part serving its purpose and serving the others. So too should a class community be a microcosm of the world.

Eighth grade is a time of blossoming individuality, which sometimes expresses itself as self-absorption and self-centeredness. To counteract this tendency, the curriculum leads the students' gaze outward. It meets the students' growing interest in the world by striving to link everything they study in the classroom to what happens in nature and in the world that human beings have created.

In their seventh-grade study of health, nutrition, and physiology, students are excited and intrigued to learn about their bodies. In eighth grade, the curriculum becomes more objective as students begin to examine food substances—sugars, starches, fats, and proteins—not just in terms of diet but more broadly.

In seventh grade, learning was mostly led or inspired by the teacher. In eighth grade, students become increasingly able to learn on their own by following their interests and delving into new subjects. In seventh grade physics and chemistry, most of the phenomena were introduced through demonstrations, and students performed one or two lab activities per week. In eighth grade, demonstrations are used only to introduce lab techniques or to demonstrate the phenomena that are too complex or dangerous for students to work with on their own. Most of the time, students benefit from working alone or in small lab groups.

In his lectures on the curriculum in *Discussions with Teachers*, Rudolf Steiner spoke about the curriculum of the upper grades as a vast channel into which various streams flow. Every subject connects with and informs every other subject. By working with such a curriculum, students develop a stronger sense of themselves and their connection to the world around them. The eighth-grade chemistry block offers many opportunities for students to deepen and broaden their perspective.

The chemistry curriculum for eighth grade focuses on the following topics:

- The nature of the major food substances: sugar, starch, fats, and proteins
- The commercial and industrial connections and applications related to these substances

For each of these substances—sugars, starches, fats, and proteins—we will examine the following:

- history
- sources
- extraction and refinement
- nutritional, domestic, commercial, and industrial uses
- physical and chemical characteristics
- interactions with water, heat, acids, and bases

We will also learn the chemical tests that allow us to identify the substance.

Such a course of study reveals the most obvious as well as the more mysterious characteristics of these common substances, and it will help students develop a new appreciation for human beings who work with these substances in so many different ways.

Scheduling Considerations

If the eighth-grade chemistry block is three weeks, rather than four, I suggest using the first two days of the final week to focus on fats, the next two days on proteins, and the last day for the final chemistry exam. Choose the topics for fats and proteins that you find most relevant and meaningful and, if you want to broaden the students' perspective, assign some activities as homework or independent projects.

The following verse by Rudolf Steiner captures the essence of our work in eighth-grade chemistry:

Seek in your own being
And you will find the world;
Seek in the world-wide being
And you will find yourself.
Note the constant swing
Between self and world,
And you will find revealed
The human cosmic being
The cosmic human being.

Week 1 of Grade 8 Chemistry
SUGAR

In eighth grade chemistry, each of the food substances will be studied in terms of its chemical interactions, but such studies will be most meaningful if the substance is also examined in the broadest possible context.

Before embarking on our study of the four major food substances, it is important to remind the students that all of these substances are a result of the most important chemical process on earth: photosynthesis. Through photosynthesis, plants use sunlight to transform water and carbon dioxide into glucose, a simple sugar. From this glucose, plants create a host of other substances: starch, cellulose, fats, and proteins. It is not hyperbolic to say that photosynthesis is the basis for life on earth.

Students should then be told about sugar's history, sources, production, and uses. Sugarcane was originally domesticated in New Guinea and in China. From there it spread westward to India and eventually to Europe, but it was so rare and expensive that it was used sparingly.

In the seventeenth century, sugar cultivation in the New World grew, and with it, the reliance on slave labor. During the Napoleonic wars, England imposed a blockade on Europe, cutting off supplies of cane sugar, but scientists found ways to extract sugar from sugar beets. Today, Brazil and North America are the world's two largest producers of sugar. Fortunately, machinery has been developed to reduce the industry's reliance on manual labor.

To grow sugar, the field must be plowed and harrowed. Then billets of cane (sections of stem that have root nodes) are planted, and the fields are watered and weeded. Although sugarcane grows quickly, it takes up to two years to mature. Before harvesting, the sugarcane field may be set on fire to burn off most of the leaves, but the stalks do not burn because of their high moisture content and tough outer shell.

The cut stalks are taken to the processing facility, where they are washed and crushed. The cane juice is clarified with lime to settle out impurities; then it is put in huge evaporator tanks that reduce the liquid until crystals begin to form in the thick syrup. The crystals are separated out by centrifuges and then refined further and dried. Except in its chemical composition, refined sugar bears little resemblance to the cane juice from which it is derived.

Sugar's primary use is in food, but not just as a sweetener. It serves as a preservative, as a bulking and texturizing agent, and as a food coloring. Sugar from many sources provides the raw materials for brewing and distilling of alcoholic beverages. Ethanol, produced from sugar, is quickly becoming a popular additive or alternative to gasoline.

Once they have been provided with this kind of background and context for their work, students should examine many types of sugar, comparing flavor, sweetness, appearance, and texture. Then they can explore sugar's interactions with water and heat.

Sugar in the plant exists in a dissolved state, and this affinity to water expresses itself by sugar's hygroscopic nature. Sugar dissolves easily, which is why it is so useful in food and beverage preparation, and students will be amazed by the amount of sugar that can be added to boiling water. When the liquid cools, the sugar solidifies and crystallizes. This property of sugar is carefully controlled by candy manufacturers, who have to cool sugar slowly and stir it continually to control crystal size and to avoid a gritty texture.

Students will see how easily sugar burns, and they will observe that as it is heated, it goes through a series of changes of color, texture, and flavor. These changes are utilized in baking and candy making. Caramel coloring is the oldest and most widely used food coloring. Caramel coloring is manufactured not only by heating sugar, but also by using acids, and students will be impressed by the extreme reaction that occurs when we add concentrated sulfuric acid to sugar.

The students' final experience for the week is to perform a test for simple sugars using Fehling's or Benedict's solutions. These solutions go through a beautiful color change in the presence of simple sugars, and the class will find it gratifying to test various substances—and will be challenged to explain why sucrose does not yield a positive result.

Exploring sugar's relationship to water, heat, and acids and learning how to test for sugar sets the stage for the rest of the block, which will investigate the chemical properties and interactions of starches, fats, and proteins.

TOPICS

- The history, production, and uses of sugar
- Sugar interactions with water
- Sugar interactions with heat
- Sugar interactions with acid
- Cooking with sugar
- Testing for sugar

CHARACTERISTICS

- Sugar comes from many different sources.
- Sugars from different sources have different degrees of sweetness.
- Sugar is highly soluble.
- Hot water can dissolve large amounts of sugar.
- Dissolved sugar crystallizes as it cools.
- When sugar is heated intensely, it combusts and produces a carbon ash.
- When sugar is heated, it goes through a series of changes of color, texture, and flavor.
- When sugar is heated intensely, it gives off water vapor and flammable gases.
- Adding potassium chlorate to sugar causes it to combust more quickly and intensely.
- Adding concentrated sulfuric acid to sugar dehydrates and chars it.
- Sugar can be used for making various kinds of candy.
- Fehling's solution or Benedict's solution can be used to test for simple (reducing) sugars.

DAY	DEMONSTRATIONS AND ACTIVITIES	CHARACTERISTICS
1	■ The History of Sugar ■ Types of Sugar	■ Sugar comes from many different sources. ■ Sugars from different sources have different degrees of sweetness.
2	■ Sugar & Water ■ Solubility of Sugar in Hot Water	■ Sugar is highly soluble. ■ Hot water can dissolve large amounts of sugar. ■ Dissolved sugar crystallizes as it cools.
3	■ Heating Sugar ■ Controlled Heating of Sugar	■ When sugar is heated intensely, it combusts and produces a carbon ash. ■ When sugar is heated, it goes through a series of changes of color, texture, and flavor.
4	■ Intense Heating of Sugar ■ Heating Sugar Mixed with Potassium Chlorate ■ The Dehydration of Sugar by Sulfuric Acid ■ Making Candy	■ When sugar is heated intensely, it gives off water vapor and flammable gases. ■ Adding potassium chlorate to sugar causes it to combust more quickly and intensely. ■ Adding concentrated sulfuric acid to sugar dehydrates and chars it. ■ Sugar can be used for making various kinds of candy.
5	■ Testing for Sugar 1 ■ Testing for Sugar 2	■ Fehling's solution or Benedict's solution can be used to test for simple (reducing) sugars.

DAY 1

PRESENTATION

1. History of Sugar

Trace the development of the sources of sugar, from the blossom (honey), to the stem (syrup & cane juice), to the root (sugar beet), to artificial sweeteners. This development shows how humanity has taken ever greater control of the substance.

Students should be told about how sugarcane was domesticated, brought to the New World, and became the basis for the slave trade. They should learn about sugarcane growing, harvesting, and refining.

The many uses of sugar can be described, and it is important to consider sugar in terms of nutrition.

RELATED TOPICS

- Sugarcane harvesting in the Caribbean and/or the American South
- Atlantic slave trade
- Sugar consumption in Western cultures
- Sugar in nutrition
- Sugar and alcohol
- Illnesses and diseases associated with high sugar consumption

ACTIVITY

2. Types of Sugar

Characteristic: Sugar comes from many different sources.

Characteristic: Sugars from different sources have different degrees of sweetness.

PROCEDURE

- Students should be given pieces of sugarcane to examine.
- Then they should be given slivers of the cane to chew and to extract the cane juice.
- Students should taste different types of sugars and artificial sweeteners and note their characteristics.

SAFETY

- Cut the sugarcane carefully; the outer layers are tough so it's easy for the knife to slip and cause an injury.
- All waste from this activity can be safely disposed of in the trash or compost.

RELATED TOPICS

- Differences between natural and artificial sweeteners
- Discovery of the beet sugar
- History of artificial sweeteners

MATERIALS

pieces of sugarcane
an assortment of natural and artificial sweeteners, such as:
honey
maple syrup
raw sugar
refined white sugar
molasses
stevia
corn syrup
saccharine
aspartame
sucralose

DAY 2

LAB ACTIVITY

3. Sugar & Water

Characteristic: Sugar is highly soluble .

EQUIPMENT

measuring cup
beam balance or scale (optional)
2 large beakers
stirring spoon or rod
electric kettle

MATERIALS

sugar
water

PROCEDURE

- Measure out a cup or 250 ml of sugar. *Note: Students can weigh the sugar if they want to work more precisely.*
- Fill a large beaker with 250 ml of cold water.
- Pour a small amount of sugar into the water and stir.
- Wait until all the sugar dissolves, then add more sugar.
- Note whether the speed of dissolution changes as more sugar is added.
- Stop adding sugar when there are solid sugar crystals on the bottom of the beaker and no more sugar dissolves.
- Measure how much sugar is left from the original 250 ml of sugar.
- Repeat this activity with water that has been heated to boiling in an electric kettle.

COMMENT

Reserve one of the groups' hot water-sugar solutions to use for the following demonstration.

DEMONSTRATION

4. The Solubility of Sugar in Hot Water

Characteristic: Hot water can dissolve large amounts of sugar.

Characteristic: Dissolved sugar crystallizes as it cools.

PROCEDURE

- **Setup**
  - Attach a ring clamp to a laboratory stand and place a wire screen on top.
  - Attach a 500-ml Erlenmeyer flask to the stand with a utility clamp so that it rests on the wire screen.
  - Set up the Bunsen burner under the ring clamp and adjust the height.

  Alternatively, use a medium-sized pot and a hot plate.
- **Process**
  - Use one of the hot water solutions of sugar, noting how much sugar has already dissolved in the 250 ml of hot water.
  - Heat the flask with a Bunsen burner or the pot with a hot plate until the solution boils.
  - Measure out another cup or beaker of sugar and weigh if desired.
  - Add sugar to the boiling solution and stir.
  - Keep adding sugar. If the solution absorbs the entire cup, add more.
  - After you finish this demonstration, pour the solution into a medium beaker and set it aside to cool.

SAFETY

- Wear goggles, gloves, and an apron for this demonstration.
- As a rule, students shouldn't taste or eat substances worked with in the lab, but an exception can be made in this case.
- If students are going to taste the sugar after it crystallizes, it is important to wash the glassware and utensils thoroughly before using them.

EQUIPMENT

Erlenmeyer flask, 500 ml
pot, medium-sized (optional)
laboratory stand
ring clamp
utility clamp
wire screen
Bunsen burner
hot plate (optional)
stirring spoon
beam balance or scale (optional)
2 medium beakers

MATERIALS

solution of water and sugar from previous activity

- The sugar-water mixture can be diluted and safely flushed down the sink.

COMMENT

Because sugar dissolves so readily, one could keep adding more sugar to the solution and it would continue to absorb it. As the solution cools, the liquid sugar will crystallize. As a possible class or homework project, students can make rock candy.

RELATED TOPICS

- How sugar's solubility is utilized in sugar refining
- Structure and properties of sugar crystals
- Use of powdered and decorative sugars in baking and confections

DAY 3

LAB ACTIVITY

5. Heating Sugar

Characteristic: When sugar is heated intensely, it combusts and produces a carbon ash.

How substances combust reveals a great deal about their chemical composition. When this is done in a controlled way, it is called qualitative analysis.

PROCEDURE

- This lab activity is best done in groups of 2 or 3 students.
- Set up the chafing dish fuel in the sand tray and light it.
- Put a small amount of sugar in the deflagration spoon and hold the spoon over the flame.
- As the sugar begins to melt around the edges, hold it higher above the flame so that the reaction proceeds more slowly. Note the speed of melting.
- Continue to heat until all the sugar melts, turns color, and begins to smoke.
- Students can try to ignite the vapors with a burning splint.
- Once the sugar is burning, the spoon can be removed from the flame and the sugar will continue burning on its own.
- Note the speed of combustion and any byproducts.
- When the combustion has stopped, examine the ash.

EQUIPMENT

sand trays
stainless steel soup spoons
pliers or tongs

MATERIALS

chafing dish fuel
matches
sugar
splints

SAFETY

- Students should wear goggles and gloves for this activity.
- Students will need instructions of how to light the chafing dish fuel, how to hold the spoon, where the spoon should be held in relation to the flame, etc.

- The chafing dish fuel can be extinguished by using a pliers or tongs to replace the lid.
- The sugar ash can be safely disposed of in the trash.

COMMENT

Stainless steel soup spoons can be used for deflagration spoons. Minor ash deposits can be cleaned with sandpaper and steel wool. Major deposits can be cleaned by heating the spoons until they are red hot and then quenching them in cold water.

The products of combustion—heat, light, smoke, water vapor, and ash—reveal the chemical components of sugar: hydrogen, oxygen, and carbon.

Students will observe that sugar contains a great deal of potential chemical energy. If they didn't learn about calories in seventh grade, this activity can serve as a foundation.

RELATED TOPICS

- The history of the calorie as a unit of measurement
- Caloric value of sugar
- Creating alcohol gel

LAB ACTIVITY

6. Controlled Heating of Sugar

Characteristic: When sugar is heated slowly, it goes through a series of changes in color, texture, odor, and flavor.

PROCEDURE

- Set up the chafing dish fuel in a sand tray.
- Using a laboratory scoop or spoon, put about ½ teaspoon of sugar in a test tube.
- Using a pipette, add two or three drops of water to the sugar.
- Light the fuel and hold the test tube with a test tube clamp over the flame at an angle, pointing away from everyone.

- Heat the test tube as gently as possible and move the test tube a little bit so that the sugar heats evenly and doesn't scorch.
- If the sugar is heated slowly enough, it will turn into a clear liquid before beginning its color changes.
- Continue to heat the sugar, noting the color changes from straw to amber to orange to brown to black.
- Once the sugar begins to smoke, light a splint and use it to light the vapors arising from the mouth of the test tube.

SAFETY

- The pipette can safely be disposed of in the trash.
- The test tube with the burned sugar should be soaked in hot water until the sugar loosens and can be cleaned out.
- The chafing dish fuel can be extinguished by using a pliers or tongs to replace the lid.

COMMENT

Students should learn the names for each of the stages the sugar goes through when it is heated.

When the students make candy the next day, they will need to keep careful track of the temperature of the sugar to insure that the candy is the correct hardness and consistency.

EQUIPMENT

test tubes
test tube racks
test tube clamps
laboratory scoops
plastic pipettes
pliers or tongs

MATERIALS

chafing dish fuel
small paper baking cups containing sugar
matches
small cup with water
splints

DAY 4

DEMONSTRATION

7. Intense Heating of Sugar

Characteristic: When sugar is heated intensely, it gives off water vapor and flammable gases.

EQUIPMENT

laboratory stand with test tube clamp

test tube

Bunsen burner

laboratory safety shield

matches or lab striker

MATERIALS

splint

PROCEDURE

- **Setup**
  - Attach the test tube clamp to the laboratory stand and adjust it so that the test tube is held at about a 45° angle.
  - Set the Bunsen burner underneath the test tube so that the flame will heat the tube intensely.
- **Process**
  - Put about 1 inch of sugar in the test tube.
  - Light the Bunsen burner and set to a hot flame.
  - Students should observe condensation on the inside of the test tube.
  - Once vapors start being released, light the splint in the Bunsen burner flame and hold it over the mouth of the test tube to see if the vapors ignite.
  - Observe how the vapors ignite and continue to burn.

SAFETY

- Goggles and gloves should be worn for this demonstration.
- This demonstration should be done behind a laboratory safety shield.

COMMENT

Sugar is a carbohydrate, which means that it is composed of hydrogen, oxygen, and carbon. On Day 3, when students burned sugar, they saw that sugar contained carbon because the ash was black. When they gently heated sugar in the test tube, the condensation of water revealed the presence of hydrogen and oxygen in sugar. In this day's demonstration the flammable gases will reveal the presence of hydrogen and carbon.

DEMONSTRATION

8. Heating Sugar Mixed with Potassium Chlorate

Characteristic: Adding potassium chlorate to sugar causes it to combust more quickly and intensely.

This demonstration gives the students an experience of enhanced combustion of sugar and should impress them with how much energy sugar contains.

PROCEDURE

- **Setup**
  - Set the laboratory stand on a sand tray so that any sugar or ash that overflows the spoon will be caught.
  - Attach the test tube clamp to the lab stand and adjust it so that it holds the deflagration spoon level.
  - Set up the Bunsen burner underneath the spoon.
  - Using separate scoops or spoons for each substance, put about one tablespoon of sugar and about ½ tablespoon of potassium chlorate on a piece of paper.
  - Tilt the paper gently back and forth to mix the substances.
  - Pour the sugar-potassium chlorate mixture into a beaker and cover it with a watch glass.
- **Process**
  - Put about 1½ teaspoons of the sugar-potassium chlorate mixture in the deflagration spoon.

EQUIPMENT

laboratory safety shield

laboratory stand with test tube clamp

sand tray

large deflagration spoon

matches or lab striker

MATERIALS

sugar

potassium chlorate

laboratory scoops or spoons

piece of paper

- Remove the beaker with the remaining sugar-potassium chlorate mixture from the demonstration area.
- Light the Bunsen burner and adjust the heat to a gentle flame.
- Heat the sugar-potassium chlorate mixture until it ignites.

SAFETY

- Read the potassium chlorate Material Safety Data Sheet before working with it.
- Wear goggles, gloves, and an apron and do the demonstration behind a laboratory safety shield.
- Do not allow the potassium chlorate to come in contact with any other substances.
- Keep the potassium chlorate container sealed except when scooping out the needed amount.
- Once you have scooped out what is needed, cap the potassium chlorate container and remove it from the demonstration area.
- Store the potassium chlorate container in a locked cabinet separate from any combustible substances.
- If you decide to repeat the demonstration, use a fresh deflagration spoon.
- The ash from this demonstration can be safely disposed of in the trash.

COMMENT

The rapid combustion of sugar reveals the large amount of potential chemical energy contained in sugar.

Because the potassium chlorate provides additional oxygen, the combustion process is more rapid and more complete.

The degree of completeness of combustion is revealed by the texture of the ash, which will be much finer and lighter than the ash produced by regular combustion.

RELATED TOPICS

- Ignition of sugar and potassium chlorate by using concentrated sulfuric acid
- The use of sugar and potassium chlorate in military mines
- The uses of other salts, like nitrates, for explosions

DEMONSTRATION

9. The Dehydration of Sugar by Sulfuric Acid

Characteristic: Adding concentrated sulfuric acid to sugar dehydrates and chars it.

Sulfuric acid is very hygroscopic, meaning that it combines with water with an exothermic reaction. In this demonstration, the sulfuric acid combines with the hydrogen and oxygen in the sugar, leaving behind the carbon.

PROCEDURE

- Fill a beaker with 100 ml of sugar.
- Holding the other beaker firmly on a stable surface, pour about 50 ml of sulfuric acid into it.
- Cover that beaker with a watch glass to reduce the fumes from the sulfuric acid.
- Pour the sulfuric acid into the sugar and stir until it is mixed.
- When the reaction is over, move the beaker with the sugar ash to a safe place to cool.

EQUIPMENT

2 medium beakers
spoon
watch glass

MATERIALS

sugar
concentrated sulfuric acid

SAFETY

- This demonstration must either be done in a fume hood or outside. If it is done outside, the students must stand upwind so that they do not breathe the noxious sulfur oxide fumes.
- Goggles, gloves, and an apron must be worn when performing this demonstration.
- A knife may be needed to remove the ash from the beaker.
- Do not take too long to stir the sulfuric acid-sugar mixture, because the reaction will begin quickly and you need to move away.
- The ash of the dehydrated sugar can be wrapped in paper towels, put in a zip lock plastic bag, and safely disposed of in the trash.
- Wear gloves and goggles to clean the beakers that contained the sulfuric acid and the sugar.

- To clean the beakers, direct a thin stream of water into them to minimize the reaction with the remaining acid.
- The rinsing water can be neutralized with baking soda and then safely flushed down the sink.

RELATED TOPICS

- The uses of sulfuric acid as an industrial chemical

COOKING ACTIVITY

10. Making Candy

Characteristic: Sugar can be used for making various kinds of candy

Students enjoy applying their knowledge of the stages of heating sugar to candy making. Caramels, toffee, or fudge are easy to make, and recipes are readily available

MATERIALS

candy thermometers
ingredients and equipment as listed in recipe

PROCEDURE

- Obtain a recipe for the candy and try it out beforehand.
- Review the steps of the recipe with the students before they begin to cook.
- Have students work in groups of no more than three.

SAFETY

- If at all possible, cooking activities should not be done in a classroom that is used for chemistry. If there is no other alternative, clean everything thoroughly before beginning the activity.

RELATED TOPICS

- Types of candy
- Harvesting of cacao and production of cocoa and chocolate
- Biographies of Milton Hershey and/or Forrest Mars
- The history of M&Ms
- Candy advertising
- The psychology of candy marketing

DAY 5

DEMONSTRATION

11. Testing for Sugar 1

Characteristic: Fehling's solution or Benedict's solution can be used to test for simple (reducing) sugars

Testing for simple sugars provides a satisfying conclusion to the first week of the main lesson block.

The indicator solution for simple sugars is either combination of Fehling's A and Fehling's B solutions or Benedict's solution.

Each step of the preparation and testing process needs to be previewed with the students in advance so that they can perform the chemical tests on their own.

PROCEDURE

- **Preparation**
  - If you are using Fehling's solution, students will have to mix it beforehand. No preparation is needed for Benedict's solution.
  - To prepare Fehling's solution, fill a test tube with 2 inches of Fehling's A solution and another test tube with 2 inches of Fehling's B solution.
  - Pour the Fehling's B into the Fehling's A and observe the precipitation.
  - Gently shake the test tube to dissolve the crystals.
  - The solution is now ready to use.
- **Testing**
  - Pour about ½ inch of the prepared Fehling's solution into a test tube.
  - Add a few drops of maple syrup or honey into the solution.
  - Light the chafing dish fuel.
  - Holding the test tube at an angle, gently heat the solution.
  - Students should observe the change of color that indicates the

EQUIPMENT

test tubes
test tube racks
test tube clamp
plastic pipette
sand tray
pliers or tongs

MATERIALS

chafing dish fuel
matches
Fehling's A and B solutions or Benedict's solution
maple syrup or honey

presence of a reducing sugar—a monosaccharide such as glucose, fructose, or lactose.

SAFETY

- Goggles and gloves should be worn for this demonstration.
- If Fehling's or Benedict's solutions come in contact with the skin, it should be flushed with plenty of water.
- Check local ordinances regarding safe disposal.
- The chafing dish fuel can be extinguished by using a pliers or tongs to replace the lid.

COMMENT

During the review on the next day, the teacher can explain in simple terms what is happening in this reaction: The blue of the Fehling's or Benedict's solutions indicates the presence of oxidized—or rusted—copper. In the presence of a simple sugar, the copper in the solution precipitates out in its un-oxidized form, which is orange.

LAB ACTIVITY

12. Testing for Sugar 2

Students should perform the chemical test for sugar on a variety of substances. To make the chemical tests more interesting, it's best if the students are not told what substances they are testing. Substances can be labelled by number and at the end of the activity, results can be compared.

When the students have completed tests on the unknown substances, they can conduct tests on items from their snack or lunch. Be sure to include refined white sugar in the tests.

It is good to include some substances, such as baking powder, that will test negative by not showing any color change.

PROCEDURE

- **Setup**
  - Prepare a series of substances to be tested.
  - Solid substances can be put into small paper baking cups and liquid substances in small portion cups.
  - Students should work in pairs.
- **Process**
  - The students should mix enough Fehling's solution to fill each test tube with about ½ inch of solution.
  - Put the chafing dish fuel in the sand tray and light it.
  - Test each substance and record whether it is a simple sugar or not.

SAFETY

- Students should wear goggles and gloves.
- If Fehling's solution comes in contact with someone's skin, flush the area with water.
- The unused solid substances and pipettes can safely be disposed of in the trash.
- The unused liquid substances can be safely flushed down the sink.
- Check local ordinances regarding safe disposal of the used and unused solutions.
- The chafing dish fuel can be extinguished by using a pliers or tongs to replace the lid.

COMMENT

During the review the next day, the class should try to figure out why sucrose tests negative.

RELATED TOPICS

- The differences between simple and complex sugars
- How chemical tests are used in medicine and forensics

EQUIPMENT

test tubes, 12 per group of students

test tube racks

test tube clamps

plastic pipettes

sand tray

pliers or tongs

MATERIALS

chafing dish fuel

Fehling's A and B solutions or Benedict's solution, enough for 12–15 tests

small portion cups with lids

small paper baking cups for dry substances

a variety of substances that contain sugar, such as:

- orange juice
- milk or powdered milk
- malt
- fructose
- glucose
- fruit

a few substances that do not contain sugar, such as:

- salt
- baking powder
- calcium hydroxide

Week 2 of Grade 8 Chemistry

STARCH

The second week of the eighth-grade chemistry block is devoted to starch. Starch is created by plants from the sugar produced by photosynthesis. It is stored in the leaves to be used by the plant for its energy needs at night or when there is little light available for photosynthesis. Plants also store starch in their roots to use as energy reserves and in seeds to nourish the sprouting plant.

Human beings have been using grains and tubers as major food sources for millennia. Around 10,000 BCE hunters and gatherers started to settle down in stable communities and to raise crops and animals. This transformation is called the first agricultural revolution.

The second agricultural revolution in the West occurred in the eighteenth century, when greater mechanization, use of crop rotation, and a greater diversity of crops led to much higher yields. It resulted in more food being raised by a smaller number of farmers.

The third agricultural revolution—or green revolution—took place in the mid-twentieth century and had a great impact both on developed and under-developed countries. Hybridization and genetic engineering of crops and the increased use of pesticides and fertilizers produced hardier, more disease-resistant plants that produced very high yields. In the 1800s, about 90 percent of people in the US lived on farms; today, about 1 percent of our population produces our country's food and surplus for export.

Wheat is the most widely grown grain in the West, and in the East, it's rice. Students may be interested to learn how these crops are grown and the kind of work they require. Wheat farming is seasonal, and its growth cycle demands intense bursts of work alternating with periods of waiting for the crop to ripen. In contrast, rice is grown throughout the year and needs constant tending. Wheat farming relies on rainfall, which is unpredictable, while rice farming utilizes carefully controlled irrigation. Raising wheat requires personal initiative; rice growing requires collaboration. These contrasts give us insight into the cultures that rely on these grains.

Maize, or corn, is America's largest crop. Maize is a descendent of a Central American plant called "teosinte," though the modern grain hardly resembles its progenitor. The seeds of teosinte were encased in hard shells and arranged in short rows on a spike that shattered to disperse its seeds. In contrast, modern maize has hundreds of kernels attached to a cob that is completely covered by soft husks, which prevents it from reproducing without human assistance.

Corn requires a great deal of cultivation and water, but it has the highest caloric yield per acre of any grain, making it the most efficient source of animal feed. In our society, we consume a great deal of corn and its derivative, high fructose corn syrup. Another corn product—corn starch—has innumerable uses. In the food industry, cornstarch is used as a binding, emulsifying, and gelling agent. It gives food mouthfeel and is used to color foods, drinks, and sauces. Cornstarch prevents food mixes from drying and caking, and it reduces adhesion in ceramic molds. Cornstarch is used as a surface sizing agent in paper production and as glue in cardboard. The cosmetic industry uses it to thicken products and the pharmaceutical industry uses it as a binder for pills and capsules. It's good to share this kind of background and context with the students so that their work with starch has greater relevance and meaning.

Following the introductory presentation, students can extract starch from potatoes, wheat, and corn. When they examine the starch with a microscope, they will see that each type of starch has a distinctive size and shape. Later in the week, when they examine thin slices of tubers, vegetables, and fruits that have been stained with iodine, they will see how the starch is stored in the plant's tissues.

Starch can be extracted because it is insoluble in cold water. Starch's insolubility allows it to be made into Oobleck, a non-Newtonian fluid with remarkable properties. It is named for a substance found in a Dr. Seuss book. When the students review their experiences with Oobleck, it is important that they understand why it behaves as it does.

When starch is mixed with hot water, it forms a gel because the starch granules absorb water until they burst. This allows starch to be used as a thickening agent for sauces and puddings. Making pudding will give the students a new appreciation for this property of starch.

When students heat starch over a flame, they will note some similarities and some differences with sugar: the starch doesn't melt, it slowly turns brown before it begins to char, it takes longer for it to ignite, and it burns more slowly and steadily than sugar.

In seventh grade, students may have experienced the combustion of dispersed lycopodium powder; if they didn't, they should experience the dramatic combustion of dispersed corn starch now. During the review the next day, they can hear about or see a video of starch explosions, which are powerful enough to destroy grain elevators and mills.

Having learned the chemical test for sugar, students should be taught how to test for starch by using iodine, which turns from orange to dark blue in the presence of starch. Using this test, they can examine various foods and other substances, like paper, to see if they contain starch.

The final experience of the week is to convert starch to sugar. Plants do this easily, and so do we with the saliva in our mouths and the digestive juices in our alimentary tract, but converting starch to sugar in the lab is the most complicated chemical procedure performed during this block. In order to do this as a lab, students will need very clear and explicit instructions. In order for them to make sense of the results, the teacher needs to make sure that she fully understands each step of the process.

The topic of cellulose could be taken up as an alternative to the conversion of starch to sugar. We mostly encounter cellulose in the woody parts of plants, though it is also found in shells, seed coats, and seed hairs. Cellulose is a polysaccharide—a complex sugar—but it is indigestible by humans. Bacteria in the digestive systems of some ruminants allow them to derive nourishment from cellulose. Cellulose can also be broken down into sugars by concentrated sulfuric acid.

In the nineteenth century, various uses for cellulose were discovered: lacquer, nail polish, plastics, photographic film, and nitrocellulose, which was used for guncotton and smokeless powder. Rayon, made from dissolved wood pulp (cellulose), was the first synthetic fabric. Rayon still is widely used today because it can be made to imitate the textures of wool, cotton, and silk.

Even more than with sugar, the study of starch can lead the class into far-ranging explorations of processes and products used in today's world. Through these kinds of explorations, students can begin to recognize that the classroom is truly a microcosm of the world.

TOPICS

- The history and uses of starch
- Extracting starch
- Starch interactions with water
- Starch interactions with heat
- Cooking with starch
- Testing for starch
- Converting starch to sugar

CHARACTERISTICS

- Starch can be extracted from tubers and grains.
- Starch is insoluble in cold water and partly soluble in hot water.
- Starch and water can be mixed to create a non-Newtonian fluid.
- When starch is heated intensely, it combusts and produces a carbon ash.
- Adding potassium chlorate to starch causes it to combust more quickly and intensely.
- Dispersing starch greatly enhances its combustion.
- Starch can be used as a thickening agent in cooking.
- Iodine can be used to test for the presence of starch.
- Many vegetables, fruits, and grains contain starch.
- Starch can be changed into simple sugars by salivary enzymes.
- Starch can be changed into simple sugars by hydrochloric acid.

DAY	DEMONSTRATIONS AND ACTIVITIES	CHARACTERISTICS
1	■ Introduction to Starch ■ Extracting Potato Starch ■ Extracting Wheat Starch ■ Extracting Corn Starch	■ Starch can be extracted from tubers and grains.
2	■ Starch and Water ■ Working with a Non-Newtonian Liquid	■ Starch is insoluble in cold water and partly soluble in hot water. ■ Starch and water can be mixed to create a non-Newtonian fluid.
3	■ Heating Starch ■ Heating Starch Mixed with Potassium Chlorate ■ The Combustibility of Dispersed Starch ■ Making Pudding	■ When starch is heated intensely, it combusts and produces a carbon ash. ■ Adding potassium chlorate to starch causes it to combust more quickly and intensely. ■ Dispersing starch greatly enhances its combustion. ■ Starch can be used as a thickening agent in cooking.
4	■ Testing for Starch 1 ■ Testing for Starch 2 ■ Testing Food Substances for Starch	■ Iodine can be used to test for the presence of starch. ■ Many vegetables, fruits, and grains contain starch.
5	■ Converting Starch to Sugar Using Salivary Enzymes ■ Converting Starch to Sugar Using Hydrochloric Acid	■ Starch can be changed into simple sugars by salivary enzymes. ■ Starch can be changed into simple sugars by hydrochloric acid.

DAY 1

PRESENTATION

1. The History of Starch

In this presentation, we can trace the evolution from hunting-gathering cultures to cultures that domesticated animals and plants. Discuss the impact of the first agricultural revolution on society and how it led to greater permanence in settlements.

Students should review what they know about the major grains and what they learned about starches and other carbohydrates in their seventh-grade study of health and nutrition.

It is also appropriate to review what they know about photosynthesis and to reiterate the importance of this process for all life on earth. This presentation should include a description of the function of starch in the life of the plant and how plants store starch. The carbon cycle can also be described.

RELATED TOPICS

- The role of starch (or carbohydrates) in nutrition
- Major grains of the world – how they are cultivated and used
- The three agricultural revolutions
- Genetically modified crops
- The use of pesticides and herbicides in modern agriculture
- Non-food-related uses of starch

DEMONSTRATION & LAB ACTIVITY

2. Extracting Potato Starch

Characteristic: Starch can be extracted from tubers and grains.

EQUIPMENT

potato peeler
large beaker
grater with coarse holes
strainer or colander
microscope
microscope slides with cover slips
watch glass

MATERIALS

russet potato
water

PROCEDURE

- Peel a potato and grate it. Put the shreds into a large beaker of water and stir.
- Observe how the water turns milky and how starch begins to collect on the bottom.
- Strain the shredded potato and pour off most of the water.
- Scrape up the layer of starch on the bottom of the beaker and allow it to dry out on a watch glass.
- If a microscope is available, examine the starch under the microscope.

SAFETY

- Potato peelings can safely be disposed of in the trash.

RELATED TOPICS

- The domestication and cultivation of the potato
- Ireland and the potato famine
- The history, production, and marketing of French fries

LAB ACTIVITY

3. Extracting Wheat Starch

The lab is best done in groups of two.

PROCEDURE

- Put a small amount—1 or 2 tablespoons—of flour into the bowl.
- Pour a little water into the flour and stir to mix.
- Add more water or flour as necessary to make a dough.
- Knead the dough in the bowl and then on the counter until it is springy, about 3 minutes. Sprinkle extra flour on the counter so the dough does not stick to it.
- Put a small beaker or jar upside down on top of the plate or large watch glass.
- Remove a small ball (about 3/4 inch in diameter) from the kneaded dough and place it on of the upside-down beaker or jar.
- While one student kneads the dough, the other uses a wash bottle or pipette to allow a thin stream of water to wash over the dough and flow onto the plate or watch glass.
- The water running off the dough ball should begin to look milky because of the starch that is mixed in the water.
- Set the plate with the starch water aside so that the water can evaporate, leaving the dry starch.
- If a microscope is available, examine the starch under the microscope.

EQUIPMENT

small beaker or jar
watch glass or plate
wash bottle
plastic pipette
microscope
microscope slides with cover slips

MATERIALS

unbleached white flour
water

SAFETY

- The starch water can be safely be flushed down the sink.
- The pipette can safely be disposed of in the trash.

COMMENT

It's important not to squirt the water too quickly or forcefully.

Students can continue kneading the dough ball until all the starch has been washed out. What will remain is a brown, stringy substance, which is the gluten in the flour.

ACTIVITY

4. Extracting Corn Starch

EQUIPMENT

blender
two medium bowls
fine mesh strainer
spatula
cheesecloth
plate
microscope
microscope slides with cover slips

MATERIALS

white hominy corn
water

PROCEDURE

- Pour about a cup of hominy corn into a bowl.
- Cover with water and allow to soak overnight.
- Pour out the water and rinse off the corn
- Put the corn into a blender and add enough water to cover.
- Blend until smooth.
- Put a fine mesh strainer over a bowl and pour the corn-water mixture through the sieve.
- Agitate the mixture in the sieve with a spatula or with your hand.
- Add water to the mixture in the sieve to thin it as it is being agitated.
- Eventually, only the husks of the corn will remain in the sieve and the bowl will contain the cornstarch-laden water.
- Clean the sieve and then line it with cheesecloth or other very fine mesh cloth and put over a bowl.
- Pour the cornstarch-water mixture through the cloth.
- Allow the mixture to sit in the bowl until most of the cornstarch settles on the bottom.
- Carefully pour off the water, scrape out the cornstarch, and put it on a plate to dry.
- If a microscope is available, examine the starch under the microscope.

SAFETY

- All the food products of this lab can be safely disposed of in the trash or compost and the liquids flushed down the sink.

COMMENT

As part of the review, have students look at potato starch, wheat starch, and corn starch under a microscope. If no microscope is available, share photographs of the starch granules.

RELATED TOPICS

- The structure of grain seeds
- How grain is milled
- Different types of wheat and their uses
- Grains in brewing and distillation
- Gluten and gluten intolerance
- Vitamins in grains
- Vitamin deficiency diseases

DAY 2

LAB ACTIVITY

5. Starch & Water

Characteristic: Starch is insoluble in cold water and partly soluble in hot water.

Unlike sugar, which mixes readily with water, starch is relatively insoluble in cold water, and only slightly soluble in hot water. Cooked starch forms a gel that traps water and is therefore used as a thickener in sauces, gravy, or puddings.

PROCEDURE

- **Solubility of Starch in Cold Water**
  - Pour about 8 ounces of very cold water into a medium beaker or jar.
  - Drop 1 tablespoon of starch onto the water and observe what happens if you don't stir.
  - Stir the water and observe whether the starch dissolves.
  - Stir the water vigorously and then allow it to stand undisturbed.
  - Observe what happens to the starch when the water has been standing for a few minutes.

EQUIPMENT

2 medium beakers
spoon or laboratory scoop
electric kettle
pot (optional)
a hot plate (optional)
small beaker or jar
test tube
test tube clamp
test tube rack
pliers or tongs
matches or butane lighter

MATERIALS

water
cornstarch
chafing dish fuel

- **Solubility of Starch in Hot Water**
  - Repeat the activity above with water that has been heated in an electric kettle or in a pot on a hot plate.
- **Creating a Gel**
  - Create a mixture of 1 part starch and 2 parts cold water in a small beaker or jar.
  - Pour 1 inch of the mixture into a test tube.
  - Put the chafing dish fuel in the sand tray and light it.
  - Using a test tube clamp, hold the test tube at an angle and gently heat the starchy water until it forms a gel.

SAFETY

- Wear goggles and gloves when heating the starch in the test tube.
- The starch-water mixtures can be safely flushed down the sink and the gel disposed of in the trash.
- The chafing dish fuel can be extinguished by using a pliers or tongs to replace the lid.

LAB ACTIVITY

6. Working with a Non-Newtonian Fluid

Characteristic: Starch and water can be mixed to create a non-Newtonian fluid.

A non-Newtonian fluid is a fluid that doesn't follow Newton's law of viscosity, which states that viscosity stays constant independent of stress.

In a non-Newtonian liquid, the viscosity can change when force is applied. In the case of Oobleck (named after a substance in a Dr. Seuss story), the liquid becomes more solid. In the case of ketchup, when the bottle is hit suddenly, the liquid becomes runnier.

Each student should make her own Oobleck.

PROCEDURE

- Making Oobleck
  - Put about ½ cup of cornstarch in a bowl.
  - Add ¼ cup of cold water, a few tablespoons at a time, stirring the starch and water mixture.
  - When the mixture begins to resist stirring and looks like it's cracking, test whether it forms a coherent mass again when it's not being stirred.
  - If so, it is ready.
  - If necessary, add very small amounts of water until the mixture resists stirring.
- **Working (Playing) with Oobleck**
  - Quickly poke a finger into the Oobleck. It should resist the finger.
  - Slowly insert the finger into the Oobleck. The finger should penetrate the Oobleck.
  - Take some Oobleck out of the bowl and put it in your palm. Close your palm quickly. Open your palm. The Oobleck will run and fall through your fingers.
  - Gather the Oobleck on a counter or plate by quickly pushing it together with your fingers.

EQUIPMENT

bowls
stirring spoons

MATERIALS

cornstarch
water
food coloring (optional)

SAFETY

- The Oobleck can be safely disposed of in the trash or diluted and flushed down the sink.

COMMENT

If students wish to color the Oobleck, drop food coloring into the water before adding it to the cornstarch.

The Oobleck leaves a film of cornstarch on any surface that it touches. This can be wiped up with a sponge or wet paper towel.

During the review the next day, students may enjoy watching a video clip of people walking, running, dancing, or biking on a tub or even a swimming pool filled with Oobleck.

RELATED TOPICS

- Viscosity
- Properties of non-Newtonian fluids
- Examples of non-Newtonian fluids:
  - Bingham plastics
  - Flubber or slime
  - Silly Putty
  - ketchup
- Uses of non-Newtonian fluids in industry

DAY 3

ACTIVITY

7. Heating Starch

Characteristic: When starch is heated intensely, it combusts and produces a carbon ash.

EQUIPMENT

sand trays
deflagration spoons
pliers or tongs
matches or butane lighter

MATERIALS

chafing dish fuel
starch

PROCEDURE

- Place chafing dish fuel in the sand tray and light it.
- Put a small amount of starch in the deflagration spoon and hold the spoon over the flame.
- Note that unlike sugar, which melts around the edges, the starch only turns brown around the edges, and as it turns into dextrin, it releases an odor like toasted bread.
- Continue to heat the starch until it begins to smoke and eventually ignites.
- Once the starch is burning, you can remove the spoon from the flame to see if the starch burns on its own.
- Note the speed of combustion and any byproducts.
- When the combustion has stopped, examine the ash.
- Extinguish the chafing dish fuel by using a pliers or tongs to replace the lid.

SAFETY

- Students should wear goggles and gloves for this activity.
- Deflagration spoons can be cleaned with sandpaper or steel wool.
- If they are heavily crusted, they can be heated until they are red hot and then quenched in a beaker of cold water.

COMMENT

Starch may not ignite as quickly as sugar, but it reveals that it contains a great deal of stored chemical energy by how well it burns. Like sugar, the products of combustion—heat, light, smoke, water vapor, and ash—reveal the chemical components of starch: hydrogen, oxygen, and carbon.

These comparisons can be developed further by burning starch mixed with potassium chlorate in the next demonstration.

RELATED TOPICS

- Starch's caloric values
- Dextrin—its production and uses
- The Maillard (browning) reaction

DEMONSTRATION

8. Heating Starch Mixed with Potassium Chlorate

Characteristic: Adding potassium chlorate to starch causes it to combust more quickly and intensely.

PROCEDURE

- **Setup**
  - Set the laboratory stand on the sand tray so that any starch or ash that overflows the spoon will be caught.

EQUIPMENT

laboratory safety shield
laboratory stand
utility clamp
sand tray
laboratory scoops or spoons
large deflagration spoon

MATERIALS

corn starch
potassium chlorate
piece of paper
lab striker or butane lighter

- Attach the test tube clamp to the lab stand and adjust it so that it holds the deflagration spoon level.
- Set up the Bunsen burner underneath the spoon.
- Using separate scoops or spoons for each substance, put about one tablespoon of starch with about ½ tablespoon of potassium chlorate on a piece of paper.
- Tilt the paper gently back and forth to mix the substances.
- Pour the starch-potassium chlorate mixture into a beaker and cover it with a watch glass.

Process

- Put about 1½ teaspoons of the starch-potassium chlorate mixture in the deflagration spoon.
- Remove the beaker with the remaining starch-potassium chlorate mixture from the demonstration area.
- Light the Bunsen burner and adjust the heat to a gentle flame.
- Heat the starch-potassium chlorate mixture until it ignites.

SAFETY

- Read the potassium chlorate Material Safety Data Sheet before working with it.
- Wear goggles, gloves, and an apron and do the demonstration behind a laboratory safety shield.
- Do not allow the potassium chlorate to come in contact with any substance other than the starch.
- Keep the potassium chlorate container sealed except to scoop out the needed amount.
- Once you have scooped out what is needed, cap the potassium chlorate container and remove it from the demonstration area.
- Store the potassium chlorate container in a locked cabinet separate from any combustible substances.
- If you decide to repeat the demonstration, use a fresh deflagration spoon.
- The ash from this demonstration can safely be disposed of in the trash.

COMMENT

It's good to have the students compare and contrast the combustion of sugar and starch in terms of:

- change of state prior to combustion
- speed of ignition
- rate of combustion
- byproducts during combustion
- ash

DEMONSTRATION

9. The Combustibility of Dispersed Starch

Characteristic: Dispersing starch greatly enhances its combustion.

If you did not do this demonstration with lycopodium powder in seventh grade, it is worth doing it with corn starch in eighth grade to illustrate the extreme combustibility of dispersed starch.

PROCEDURE

- Attach the plastic tubing to the funnel.
- Put a small piece of wire screen and a small piece of tissue paper in the narrow bottom of the funnel so that the starch cannot fall into the tube.
- Light the Bunsen burner and adjust the flame so that it is a high yellowish flame.
- Carefully pour about a teaspoon of corn starch into the funnel.
- Kneel down so that you are below the position of the Bunsen burner, but can hold the funnel upright close to the top of the barrel of the Bunsen burner.
- Hold the free end of the plastic tubing and bring it to your lips.
- With a quick, hard exhalation, blow the starch into the flame.
- As soon as the starch has combusted, turn off the Bunsen burner.

EQUIPMENT

funnel

plastic tubing that fits on the end of the funnel, 2′ long

small piece of wire screen

small piece of tissue paper

Bunsen burner

spoon or laboratory scoop

butane lighter or lab striker

MATERIALS

corn starch

SAFETY

- Make sure that the students are positioned well back from the radius of the flame column.
- The piece of tissue paper may ignite and float to the ground. Be prepared to extinguish it.
- To clean up, disconnect the funnel from the hose and wash the funnel. Blow through the hose to clear away any remaining starch.

COMMENT

Because the ignition of the starch is so dramatic, students will probably insist on seeing the demonstration repeated. I do not recommend doing it more than twice, though.

During the review, students should be told about explosions in coal mines, grain elevators, mills, and other enclosed places where starches collect. You may wish to show a video of a dramatic starch explosion.

COOKING ACTIVITY

10. Making Pudding

Characteristic: Starch can be used as a thickening agent in cooking

As students have seen, starch can be used to thicken liquids. It is very satisfying for them to utilize this property to make pudding.

PROCEDURE

(Recipe adapted from the website of America's Test Kitchen.)

- Heat 2½ cups milk and all of the sugar and salt in a large saucepan over medium heat until simmering, stirring occasionally to dissolve the sugar.
- Meanwhile, whisk the cornstarch and another ¼ cup milk in a large bowl until no lumps remain, about 15 seconds.

- Whisk in the three egg yolks until fully incorporated, about 30 seconds.
- When the milk mixture comes to a simmer, remove it from the heat and, whisking constantly, slowly add the milk mixture to the yolk mixture to temper the eggs. Pouring the milk slowly and whisking constantly will prevent the eggs from cooking too fast.
- Return the milk-yolk mixture to the saucepan.
- Return the saucepan to medium heat and cook, whisking constantly, until the pudding is thickened and registers 180°F in several places, about 1 minute.
- Taking the pan off the heat, whisk in butter and vanilla. Strain through a fine-mesh strainer set over a clean bowl.
- Spray a piece of parchment paper with vegetable oil spray and press it flush to surface of the pudding to prevent a skin from forming.
- Refrigerate until cold and set, at least 3 hours.
- Just before serving, whisk pudding until smooth.

SAFETY

If at all possible, cooking activities should not be done in a classroom that is used for chemistry. If there is no other alternative, do a thorough cleaning of all equipment and surfaces before beginning the cooking activity.

RELATED TOPICS

- Other types of starch thickeners, such as arrowroot and tapioca
- Uses of starch in cooking
- How starches thicken liquids

COMMENT

Students should work in groups of two or three for this activity. Different groups can follow different recipes.

EQUIPMENT

large saucepan
stirring spoon
whisk
large bowl
fine-mesh strainer
spatula

MATERIALS

parchment paper
vegetable oil spray
2¾ cups whole milk
½ cup (3½ ounces) sugar
¼ teaspoon salt
¼ cup (1 ounce) cornstarch
3 large egg yolks
2 tablespoons cold unsalted butter
1 tablespoon vanilla extract

DAY 4

DEMONSTRATION

11. Testing for Starch

Characteristic: Tincture of iodine turns dark blue or purple in the presence of starch.

EQUIPMENT

2 large beakers or jars
water
tablespoons

MATERIALS

cornstarch
tincture of iodine

PROCEDURE

- **Test 1**
  - Fill a beaker or jar ¾ full of cold water.
  - Sprinkle about a tablespoon of cornstarch on the surface of the water.
  - Add 7 to 10 drops of iodine to the water and starch mixture.
  - Observe the color change.
  - Stir the water and observe the final color of the water.
- **Test 2**
  - Fill a large beaker or jar ¾ full of cold water.
  - Add 7 to 10 drops of iodine to the water and stir.
  - Sprinkle about a tablespoon of cornstarch on the surface of the water. Observe the color change.
  - Sprinkle more starch onto the water.
  - Stir the water and observe the final color of the water.

SAFETY

- Wear goggles, gloves, and an apron for this demonstration.
- The solutions containing tincture can be diluted and safely flushed down the sink.

RELATED TOPICS

- Iodine—its sources and uses
- Diseases caused by iodine deficiency
- Iodizing salt

LAB ACTIVITY

12. Testing Food Substances for Starch

Characteristic: Many vegetables, fruits, and grains contain starch.

PROCEDURE

- Test a potato, radish, carrot, and apple for starch by putting a drop or two of iodine on a cut surface of the vegetable or fruit.
- *Note: the presence of starch may be subtle when viewed with the naked eye.*
- Cut very thin slices of the vegetables and fruit, put a drop of iodine on them, and view them with a magnifying glass or microscope.

SAFETY

- Students should wear goggles and gloves for this activity.
- If the tincture of iodine comes in contact with the skin, it should be flushed with water.
- The fruits and vegetables that have been tested with iodine can safely be disposed of in the trash.

EQUIPMENT

watch glasses
magnifying glass
microscope
microscope slides
coverslips

MATERIALS

a variety of substances to be tested for starch, such as potato, radish, carrot, and apple
tincture of iodine

DAY 5

LAB ACTIVITY

13. Converting Starch to Sugar Using Salivary Enzymes

Characteristic: Starch can be changed into simple sugars by salivary enzymes.

Plants create sugar by photosynthesis and convert it to starch for storage. They are then able to convert the starch back to sugar when they need sugar to support their metabolism.

In our digestion, our saliva and stomach acids allow us to convert starch into sugar, which can then be absorbed through the small intestine.

The conversion of starch to sugar is the basis for high-fructose corn syrup, as well as the production of ethanol.

EQUIPMENT

test tubes
test tube racks
test tube clamp
pliers or tongs

MATERIALS

chafing dish fuel
matches
cracker that does not contain sugar
Either Fehling's A solution and Fehling's B solution or Benedict's solution

PROCEDURE

- Each student should chew a cracker and note whether it begins to taste sweet as it is chewed.
- Spit the chewed cracker mixed with saliva into a small beaker or jar.
- If using Fehling's solution:
  - Pour 3 inches of Fehling's A solution into a test tube.
  - Repeat for Fehling's B solution.
  - Pour the Fehling's B solution into the Fehling's A solution.
  - Pour ½ inch of the mixed testing solution into a test tube.
  - Repeat with another test tube.
  - Reserve the rest of the mixed solution for other tests.
- If using Benedict's solution:
  - Pour about ½ inch of solution into each of two test tubes.
- Drop a small piece of unchewed cracker in the testing solution.
- Gently heat the solution and observe any color changes.
- If there is no color change, then the cracker does not contain sugar.

- Put a pea-sized portion of chewed cracker into the test tube with the testing solution.
- Heat gently and observe whether there is a color change.
- If the Fehling's solution turns copper color, then the saliva has converted some of the starch in the cracker to sugar.

SAFETY

- The chafing dish fuel can be extinguished by using a pliers or tongs to replace the lid
- Goggles and gloves are needed when testing for sugar.
- If Fehling's solution gets on skin, it should be flushed with plenty of water.
- Check local ordinances for safe disposal of solutions.

COMMENT

If students find it too disgusting to spit out the chewed cracker, they can use some of the cracker saliva mixture made by another student who is less squeamish.

LAB ACTIVITY

14. Converting Starch to Sugar Using Hydrochloric Acid

Characteristic: Starch can be changed into simple sugars by hydrochloric acid.

Converting starch to sugar with hydrochloric acid is a multi-step process that is challenging.

Because the lab has many steps, it is important to hand out written instructions and to review them before the students begin their work.

If the class does not seem ready for the challenge, the teacher can do this procedure as a demonstration. The teacher can then make sure the students understand each step clearly and can review what the step revealed before moving on to the next step.

EQUIPMENT

test tubes
test tube racks
test tube clamp
laboratory scoop
plastic pipette
pliers or tongs

MATERIALS

chafing dish fuel
matches
hydrochloric acid, concentrated
corn starch

PROCEDURE

- **Testing for starch**
  - Pour ½ inch of the Fehling's testing solution prepared in the previous lab activity into a test tube.
  - Using the scoop, put about ¼ teaspoon of corn starch into the solution.
  - Hold the test tube over the flame at an angle to heat the solution gently and observe any color changes.
  - If the color of the solution is unchanged, then the starch does not contain simple sugar.
- **Creating a gel**
  - In a small beaker or bowl, mix a tablespoon of corn starch with a tablespoon of water.
  - Slowly add more water and stir until it is the consistency of thin cream.
  - Pour about ½ teaspoon of the starch-water mixture into a test tube.
  - Gently heat the mixture until it begins to turn into a gel.
  - Remove from heat.
  - Extract a bit of the gel with a scoop and place it on a watch glass. This will be tested for sugar first. The rest of the gel should remain in the test tube to be used in a later step.
- **Testing the gel for sugar**
  - Pour ½ inch of the prepared Fehling's testing solution into a test tube.
  - Put the small portion of the gel that was put on the watch glass into the solution.
  - Heat the solution gently and observe any color changes.
  - If the color of the solution is unchanged, then the gel does not contain simple sugar.
- **Testing hydrochloric acid for sugar**
  - Pour ½ inch of the prepared Fehling's testing solution into a test tube.
  - Using the pipette, place 3 drops of hydrochloric acid into the solution.
  - Heat the solution gently and observe any color changes.
  - If the solution turns light blue or cyan, it indicates the presence of the acid but does not indicate the presence of simple sugar.

- **Cooking the gel with hydrochloric acid**
  - Put 4 drops of hydrochloric acid into the test tube that has the remaining starch gel.
  - Holding the test tube at a slant, heat gently and move the test tube a little until the gel dissolves. If the gel doesn't dissolve, add another drop of acid and heat. If it still doesn't dissolve, add another drop.
- **Testing the cooked gel-acid mixture for sugar**
  - Pour ½ inch of the prepared Fehling's testing solution into a test tube.
  - Carefully pour about ½ teaspoon of the dissolved acid-gel mixture into the test tube.
  - Heat the solution gently and observe any color changes.
  - If it turns copper color, it indicates the presence of simple sugar.
  - *Note: if the solution turns light blue or cyan, it shows that too much acid was added. It's likely that one or several of the lab groups will have a positive result, though.*

SAFETY

- The chafing dish fuel can be extinguished by using a pliers or tongs to replace the lid.
- Goggles and gloves must be worn for this activity.
- A minimum of acid should be dispensed and the beaker with the acid should be covered with a watch glass and put aside after the acid is dispensed.
- If Fehling's solution gets on skin, it should be flushed with plenty of water.
- To dispose, hydrochloric acid can be diluted, neutralized with baking soda or sodium hydroxide, and flushed down the sink.
- Check local ordinances regarding safe disposal of the Fehling's or Benedict's solutions.
- The pipettes can safely be disposed of in the trash.

COMMENT

Teachers should practice this activity several times, testing the quantities for accuracy and adjusting as necessary.

Week 3 of Grade 8 Chemistry

FATS & OILS

During the third week, we will explore the nature of fats and oils, which are two forms of the same substance, with fats being the solid form and oils being the liquid form. Like all plant substances, fats are made from the sugars produced in photosynthesis. The plant produces fats in order to store energy, especially for its seeds, which need nourishment until the sprouted seed can begin to photosynthesize.

Animals and humans synthesize fats from the plants and animals they eat and use these fats for insulation, energy storage and regulation, and organ protection. Fats are also a component of cell membranes. They store essential vitamins and are involved in chemical reactions regulating growth and immune function.

Vegetable oils have been used since ancient times for oil lamps, and animal fats have been used to make candles. In the seventeenth to nineteenth centuries, whale and sperm oil became the most sought-after lighting fuel in the United States. In 1850, over 350,000 barrels of whale oil were produced. Forty years later, total production was under 6,000 barrels per year because whale oil had been supplanted by kerosene and lamp oil, which were derived from petroleum.

If students have not yet learned about petroleum creation, extraction, refinement, and use, it would be good for them to do so in eighth grade. Petroleum is created from the remains of plants and algae that were subjected to heat and to tremendous pressure over millions of years. For the past century, it has served as the world's major fuel source, and its byproducts are used by numerous industries.

Plant and animal oils are essential for our diet, and they are used extensively in cooking and by the food industry. Oils taste good and have a pleasant mouthfeel; they enhance flavors and blend them together, and they contribute to the tenderness of baked goods. Oils are often used as a medium for cooking because they can get much hotter than boiling water, and they give fried foods a distinctive crispy crust.

Another major historical use of fats was to make soap. Nowadays, fats are used for lubrication of all kinds of machinery and in the cosmetic and pharmaceutical industries for ointments, creams, and lotions. Alfred Nobel's invention, dynamite—which used nitroglycerine made from oil—transformed the construction and demolition industries and was extensively used for munitions. Recently, the most promising use of plant oils is as biodiesel, which has the advantage of being a renewable fuel. With this kind of background and context, students are ready to begin to explore the properties of fats and oils.

When we examined sugar and starch, we began by investigating their relationship to water. With fats, we will begin with fire, because fats have such a strong affinity to heat. In contrast to the combustion of carbohydrates, which was incomplete and created a carbon ash, the combustion of fats is often complete, and no ash is produced. When we vaporize a candle and burn the flammable gases, we have a clear illustration that it is the gas—not the liquid or solid—that burns.

The old adage "Oil and water don't mix" points to a characteristic that is the result of these two substances' molecular makeup. Although the students don't need to know the details, it is important for the teacher to understand the differences between a polar liquid such as water and a non-polar liquid such as oil and how these affect mixing and dissolution.

Water is composed of three atoms—two hydrogen and one oxygen. The oxygen atom has a negative electrical charge, while the two hydrogen atoms have positive charges. This allows water to cohere to itself and to other substances that have polar molecules. Substances that dissolve in water are called hydrophilic. Oil is non-polar, which means that the molecule does not have a negative or positive charge. Oil can mix with other non-polar fluids, but because it does not mix with water, it is called hydrophobic.

When we shake oil and water vigorously, the oil is broken into tiny droplets that may create a suspension, but it will eventually separate. When we add a little bit of soap, however, the oil and water combine. It is enough for the students just to observe this and to hear about other types of substances that have the same characteristics, but teachers should understand the basic principles: Each soap molecule has a hydrophilic and a hydrophobic end. The hydrophobic end is attracted to the oil and unites with it, leaving the hydrophilic end of the molecule available to unite with the water. The soap serves as a chemical "bridge" that brings the two substances together, thus making the soapy oil soluble in water.

Students will be gratified to put these principles to use by making emulsions such as salad dressing or mayonnaise and by making soap. If these activities are too time-consuming, they can make candles or oil lamps, which will show them why oils were so easily used for illumination.

As students learn the most salient characteristics of fats, their consideration of its role in our bodies and diet and its myriad commercial and industrial uses will deepen their appreciation for this substance.

TOPICS

- The history and uses of fats
- Fat interactions with heat
- Fat interactions with water
- Emulsions
- Practical uses of fats

CHARACTERISTICS

- Fats and oils can be extracted from seeds.
- Fats and oils can be extracted from animal tissues.
- Fats and oils feel greasy.
- Fats and oils are less dense than water.
- Fats and oils make a translucent spot on paper.
- Paraffin melts and burns when subjected to heat.
- Paraffin and other oils combust completely and produce no ash.
- When water is dropped into burning oil, it causes the oil to flare up.
- The vapors of heated paraffin are combustible.
- Oil is not soluble in cold or hot water.
- Detergent allows partial mixing of oil and water.
- Oil and polar liquids can form an emulsion by the addition of an emulsifying agent.
- Soap can be made from oil and sodium hydroxide.
- Oil can travel through a wick and can be ignited to produce a controlled flame.

DAY	DEMONSTRATIONS AND ACTIVITIES	CHARACTERISTICS
1	■ The History & Use of Fats & Oils ■ Sources and Extraction of Fats & Oils	■ Fats and oils can be extracted from seeds. ■ Fats and oils can be extracted from animal tissues. ■ Fats and oils feel greasy. ■ Fats and oils are less dense than water. ■ Fats and oils make a translucent spot on paper.
2	■ Heating Paraffin ■ Heating Oil 1 ■ Heating Oil 2 ■ Heating Oil 3	■ Paraffin melts and burns when subjected to heat. ■ Paraffin and other oils combust completely and produce no ash. ■ When water is dropped into burning oil, it causes the oil to flare up.
3	■ The Wickless Candle ■ Oil & Water ■ Oil, Water, & Detergent	■ The vapors of heated paraffin are combustible. ■ Oil is not soluble in cold or hot water. ■ Detergent allows partial mixing of oil and water.
4	■ Emulsions ■ Emulsions in Cooking	■ Oil and polar liquids can form an emulsion by the addition of an emulsifying agent.
5	■ Making Soap ■ Making an Oil Lamp	■ Soap can be made from oil and sodium hydroxide. ■ Oil can travel through a wick and can be ignited to produce a controlled flame.

A note on scheduling: If you have scheduled four weeks for the chemistry block, you can spend this whole week on fats and oils. If you have scheduled only three weeks, use the first two days to focus on fats and oils and then two days to examine proteins. In that case, you will need to choose those demonstrations and activities that seem most important and meaningful to you. Whatever your schedule, reserve the last day of the block for the chemistry lab final exam, because it is a very satisfying way to end the block.

DAY 1

PRESENTATION

1. The History & Use of Fats & Oils

DEMONSTRATION

2. Sources, Properties, & Extraction of Fats & Oils

- Characteristic: Fats and oils can be extracted from seeds.
- Characteristic: Fats and oils can be extracted from animal tissues.
- Characteristic: Fats and oils feel greasy.
- Characteristic: Fats and oils are less dense than water.
- Characteristic: Fats and oils make a translucent spot on paper.

PROCEDURE

- **Sources and types of fats and oils.**
  - Have students examine various types of fats and oils in terms of their physical properties: state (solid or liquid), viscosity, and density. They can record their observations in a chart.
  - They will notice that fats and oils feel greasy.
  - Because fats and oils are less dense than water, they float on the water's surface.
  - One way to test for fats and oils is to see whether they create a translucent oil spot on a piece of paper.

- **Extracting oil from bacon**
  - Gently warm a frying pan.
  - Lay the bacon so that the strips are not overlapping.
  - Gently cook the bacon, turning frequently, until the fat has been rendered and the protein is browned.
  - Drain the bacon on paper towels.
  - Pour off the fat and allow it to cool.
- **Extracting tallow from beef**
  - There are many recipes available online. Some use a slow-cooker; others can be done on a stove.
  - All the recipes take many hours, but the process helps students recognize how long and hard people used to work to obtain the fat they needed to make soap.

COMMENT

There are several chemical tests for fats and oils, but the translucency test is the easiest to perform and is definitive in its results.

If you do render tallow, students can make candles from it as their final project.

RELATED TOPICS

- Saturated and non-saturated fats
- Hydrogenation of fats
- How butter is made
- Margarine and butter substitutes
- Role of fats in our diet
- Health effects of high fat consumption
- Cholesterol
- Essential oils
- Aromatherapy

EQUIPMENT

frying pan
cooking tongs

MATERIALS

various types of fats and oils
olive
coconut
corn
peanut
canola
sesame
butter
lard
tallow
mineral oil
bacon

DAY 2

LAB ACTIVITY

3. Heating Paraffin

- Characteristic: Paraffin melts and burns when subjected to heat.
- Characteristic: Paraffin and other oils combust completely and produce no ash.

EQUIPMENT

deflagration spoon
beaker, small
pliers or tongs

MATERIALS

chafing dish fuel
candle
cold water
matches

PROCEDURE

- The chafing dish fuel should be put in a sand tray and lit.
- Place a small piece of candle wax (a disk about $^1/_8$ inch thick with the wick removed) in the deflagration spoon and heat gently.
- When the paraffin has melted, pour it into a small beaker, jar, or bowl half filled with cold water and observe the change of state.
- Repeat, but allow the paraffin to ignite and to combust completely.
- Students should observe how the melted paraffin boils, releases vapors, and burns, leaving no ash.

SAFETY

- Students should wear goggles and gloves for this activity.
- Students should try to hold the spoon as steadily as possible, especially when pouring the melted paraffin.
- If paraffin spills while burning, there is no cause for alarm; it will just soak into the sand.
- The chafing dish fuel can be extinguished by using a pliers or tongs to replace the lid.

COMMENT

During the review on the following day, you may wish to demonstrate "The Wickless Candle." (See page 165.) That will help students understand that it is the paraffin vapor that is burning, not the solid or liquid paraffin.

RELATED TOPICS

- Petroleum—origin, extraction, refinement, and uses
- Non-organic oils—types, uses in industry, manufacturing, and diet
- How laxatives made of mineral oils work

LAB ACTIVITY

4. Heating Oil 1

Characteristic: Paraffin and other oils combust completely and produce no ash.

PROCEDURE

- The chafing dish fuel should be placed in a sand tray.
- Pour about ½ tablespoon of oil in the deflagration spoon.
- Light the chafing dish fuel, heat the oil, and observe how it boils.
- When the boiling oil begins to smoke, try to light the vapors by using a lit splint.
- Once the oil starts burning, move it away from the flame.
- Allow the oil to combust completely.
- Note the absence of ash.

SAFETY

- Students should wear goggles and gloves for this activity.
- Tell students to try to hold the spoon as steadily as possible.
- If oil spills while it is burning, there is no cause for alarm; it will just soak into the sand.
- The chafing dish fuel can be extinguished by using a pliers or tongs to replace the lid.

EQUIPMENT

sand tray
deflagration spoon
pliers or tongs

MATERIALS

chafing dish fuel
vegetable oil
matches
wooden splint

DEMONSTRATION

5. Heating Oil 2

EQUIPMENT
baking pan
small ladle
Bunsen burner
lab striker or butane lighter

MATERIALS
water
vegetable oil

PROCEDURE

- **Setup**
  - Fill a large baking pan with water to a depth of 3/4 inch and place it on the floor. (The oil will splatter on the floor, but it is easy to clean it up with soapy water.)
  - Set up a laboratory stand with a ring clamp that can hold a small ladle nestled in its opening.
- **Process**
  - Fill the ladle about half full of oil.
  - Heat the oil with the Bunsen burner until it begins to burn.
  - Very carefully lift the ladle out of the ring clamp and hold it at arm's length, about four feet above the baking pan.
  - Tilt the ladle to pour some of the oil into the water.
  - The burning oil will flare up when it hits the water.
  - Repeat with the rest of the oil.

SAFETY

- Goggles, heavy heat-proof gloves, and an apron must be worn for this demonstration.
- The students must stand well back from the burning oil.
- Practice this demonstration until you feel totally comfortable pouring the oil.
- The flaring oil looks dramatic and dangerous, but it behaves predictably.
- Have a fire extinguisher on hand.
- Any remaining oil can be mixed with water and soap and safely flushed down the sink.

COMMENT

In the review, see if students can figure out what is making the oil flare up when it strikes the water. Students should be told about the proper ways to extinguish oil fires.

RELATED TOPICS

- Boiling oil as a weapon
- Oil spills—causes, environmental impact, containment, remediation
- Burning oil on water

DEMONSTRATION

6. Heating Oil 3

Characteristic: When water is dropped into burning oil, it causes the oil to flare up.

This is an even more dramatic demonstration than the previous one, showing the effects of water on burning oil.

PROCEDURE

- **Setup**
  - Set up the laboratory stand with a clamp that can hold a small ladle or large deflagration spoon horizontally.
  - Raise or lower the clamp so that the flame of the Bunsen burner is the right height to heat the ladle.
  - Place the stand in a place where it will not be affected by the wind.
  - Set the large beaker with water in the spot where you'll be standing, about 10 feet from the laboratory stand.
  - Fill the syringe with water.
- **Procedure**
  - Fill the ladle about 3/4 full of oil.
  - Light the Bunsen burner and stand back about 10 feet.
  - Ask the students to be patient, because it takes a while for the oil to boil and begin to burn.

EQUIPMENT

laboratory
utility clamp
large beaker
small ladle or large deflagration spoon
Bunsen burner
lab striker or butane lighter
syringe, medium or large

MATERIALS

vegetable oil
water

- When the oil is burning, use the syringe to shoot a stream of water into the oil.
- Repeat, refilling the syringe if necessary, until all the oil has been ejected from the ladle.

SAFETY

- This demonstration must be done outside.
- The teacher must wear goggles, gloves, and an apron, and the students must stand at least 15 feet from the burning oil.
- Practice shooting a stream of water into an empty ladle so that you can perform this demonstration confidently.
- Any remaining oil can be mixed with water and soap and safely flushed down the sink.

COMMENT

In the review, see if the students can figure out why the oil flared up so dramatically. In their eighth-grade physics block, they should have learned that when water boils, the volume of steam is 1,000 times greater than the volume of water. The transformation of water into steam caused by the heat of the oil disperses the oil into fine droplets, which ignite quickly.

Make sure that students understand fully that putting out a cooking oil fire by pouring water on it is extremely dangerous. They should be told the proper ways to put out such fires.

Students might enjoy watching a video of this demonstration done on a larger scale.

RELATED TOPICS

- Classes of fires
- Types of fire extinguishers for each class of substance

DAY 3

DEMONSTRATION

7. The Wickless Candle

Characteristic: The vapors of heated paraffin are combustible.

This demonstration may help students to understand the process of combustion even better. It can be done during the review of Day 2's demonstration of burning paraffin.

PROCEDURE

- Set up the laboratory stand with a test tube clamp so that the test tube is held at a slight slant just above the flame of the Bunsen burner.
- Put a small piece of candle (¼-inch disk with the wick removed) into the test tube and heat it with the Bunsen burner with a fairly hot flame.
- Once the paraffin begins to boil and to emit vapors, try to light the vapors.
- Once the vapors ignite, turn down the Bunsen burner flame.

SAFETY

- Goggles, gloves, and an apron must be worn during this demonstration.
- Students should stand behind a laboratory safety shield.

EQUIPMENT

laboratory safety shield
laboratory stand
test tube clamp
Bunsen burner
lab striker or butane lighter

MATERIALS

candle
wooden splint

LAB ACTIVITY

8. Oil & Water

Characteristic: Oil is not soluble in cold or hot water.

EQUIPMENT

test tubes
test tube stands
test tube stoppers (optional)
beakers, small
plastic pipette
electric kettle

MATERIALS

cold water
hot water
vegetable oil

PROCEDURE

- **Oil and cold water**
  - Fill a test tube a third to a half full of cold tap water or water that has been refrigerated.
  - Add about 10 drops of oil.
  - Cap the test tube with a thumb and shake vigorously.
  - Place the test tube in the stand and observe the oil-water mixture for several minutes to observe the changes in the mixture.
- **Oil and hot water**
  - Repeat the instructions above but with water heated in the electric kettle.
  - Compare the oil-water mixtures or suspensions.
  - Save these mixtures for comparison in the next activity.

SAFETY

- If students are reluctant to cap the test tube with their thumbs, they can use stoppers.
- Students should wash their hands after these activities.
- The oil-water mixture can be mixed with soap and safely flushed down the sink.

LAB ACTIVITY

9. Oil, Water, & Detergent

Characteristic: Detergent allows partial mixing of oil and water.

PROCEDURE

- As in the last activity, create an oil-water mixture in two test tubes—one with oil and cold water and one with oil and hot water.
- Place 3 drops of detergent in each one.
- Cap the test tubes with a thumb and shake vigorously.
- Place the test tubes in the test tube stand and observe the oil-water-detergent mixture for several minutes.
- Compare the four test tubes (from the previous activity and this one) throughout the day and on the next day prior to the review.

SAFETY

- If students are reluctant to cap the test tube with their thumbs, they can use stoppers.
- The oil-water-detergent mixtures can be mixed with soap and safely be flushed down the sink.
- The pipettes can safely be disposed of in the trash

COMMENT

- Water and oil are good examples of polar and non-polar liquids. The students will benefit from hearing a simple explanation of atomic theory prior to going to high school, and the nature of oil and water provides the perfect opportunity.
- Students can be told how the molecular nature of soap allows it to bond to both the polar water molecules and the non-polar oil molecule.

RELATED TOPICS

- Difference between soaps and detergents
- "Green" detergents
- Other examples of polar and non-polar liquids
- How dry-cleaning laundry works

EQUIPMENT

test tubes
test tube stands
plastic pipettes
electric kettle
beakers, small

MATERIALS

cold water
vegetable oil
hot water

DAY 4

PRESENTATION

10. Emulsions

Building on the experience with oil, water, and detergent, the teacher can introduce students to emulsions. An emulsion has droplets of one liquid evenly dispersed within another liquid. This makes the emulsion thicker than the original liquids.

A temporary emulsion will separate after a few hours; a semi-permanent emulsion will last longer, but will eventually separate, and a permanent emulsion will stay unified for an extended period of time.

A salad dressing made from oil and vinegar is a temporary emulsion; a sauce thickened with cornstarch is a semi-permanent emulsion, and mayonnaise is a permanent emulsion.

Milk is an emulsion of water, protein solids, and butterfat. Students will see how these components separate when they curdle milk the following week.

COOKING ACTIVITIES

11. Emulsions in Cooking

Characteristic: Oil and polar liquids can form an emulsion by the addition of an emulsifying agent.

MATERIALS

recipes for salad dressing and gravy

ingredients and equipment to make the recipes

PROCEDURE

- **Salad dressing**
  - Students can make salad dressing, using mustard as an emulsifying agent.
- **Sauce or gravy**
  - Students can make a sauce or gravy thickened with cornstarch, which serves as the emulsifying agent.

- **Mayonnaise**
  - Mayonnaise uses egg yolks as the emulsifying agent.
  - In a food processor, combine yolks, lemon juice, mustard, sugar, salt, and pepper until combined, about 10 seconds.
  - With the machine running, gradually add oil in a slow steady stream. The process should take about 30 seconds.
  - Scrape down the sides of the bowl with a rubber spatula and process 5 seconds longer.
  - Adjust seasoning with additional salt and pepper, if needed.

SAFETY

- If at all possible, cooking activities should not be done in a classroom that is used for chemistry. If the classroom must be used, surfaces, equipment, and utensils need to be washed thoroughly beforehand.
- Students should work in lab groups of two or three.

COMMENT

Mayonnaise can also be made with an immersion blender.

EQUIPMENT

food processor
spatula

MATERIALS

2 large, pasteurized egg yolks
4 tsp. fresh lemon juice
¼ tsp. Dijon mustard
⅛ tsp. sugar
¼ tsp. table salt
¾ cup canola or other vegetable oil

DAY 5

LAB ACTIVITY OR DEMONSTRATION

12. Making Soap

Characteristic: Soap can be made from oil and sodium hydroxide.

Making soap is a wonderful culmination of the study of fats and oils. This remarkable substance can be made from just two ingredients: fat and lye.

There are many soap recipes available, some quite simple and others much more complicated. I leave it to teachers to find ones that are most suitable.

MATERIALS

recipes for soap

ingredients and equipment to make the recipe

PROCEDURE

Follow the chosen recipe.

SAFETY

- Everyone should wear goggles, gloves, and aprons when working with sodium hydroxide.
- Any leftover sodium hydroxide can be diluted, neutralized with vinegar, and safely flushed down the sink.

COMMENT

Household lye works just as well as sodium hydroxide purchased from a chemical supply company.

Hard soap needs time to cure, which can be frustrating for the students, so if you are making hard soap, reserve a little in its soft stage to use right away.

DAY 5 (ALTERNATIVE)

LAB ACTIVITY

13. Making an Oil Lamp

Characteristic: Oil can travel through a wick and can be ignited to produce a controlled flame.

Olive oil lamps were used throughout the ancient world. They were typically shallow dishes made of clay with a wick of linen.

PROCEDURE

- Cut the wire to a 15-inch length.
- Wind one end of the wire five or six times around the nose of the pliers to create a spiral that is narrower at the top and wider at the bottom.
- Coil the rest of the wire into a circle that fits snugly into the bottom of the jar.
- Adjust the spiral so that it will stand upright and hold the wick just above the surface of the oil.
- Feed the wick or the cloth through the spiral from the bottom and pull it through the top.
- Place the spiral with the wick in the jar and fill the jar with oil to about half an inch below the top of the wick.
- Allow the wick to absorb olive oil to the top of the wick
- Once the wick is saturated, light the wick.

SAFETY

- To extinguish the oil lamp, use a candle snuffer.
- The jar and the oil won't heat up from the olive oil flame, but it's best to keep it on a nonflammable surface as a precaution.
- The wick needs to be trimmed before relighting the lamp. To do that, cut off the charred end to create a clean edge.

EQUIPMENT

Needle-nose pliers

MATERIALS

matches

wire, 16-, 18-, or 20-gauge

braided wick or piece of cotton or linen cloth, 2″ x 5″

4-oz. jelly jar

olive oil or vegetable oil

COMMENT

A simpler version of this oil lamp can be made with a jar and lid. Make a hole in the lid for the wick. Fill the jar with oil. Insert the wick in the lid's hole and screw the lid on. Be sure that the wick reaches the oil. This type of lamp does not need the wire wick holder.

The very simplest oil lamp can be made with a wick of string or rolled-up cotton draped over the edge of a jar or bowl filled with oil.

Alternatively, students can make candles as an independent project at home. Procedures are readily available in books or online recipes.

RELATED TOPICS

- Types of oil lamps used in the ancient world
- The development of the lantern
- The whaling industry

Week 4 of Grade 8 Chemistry

PROTEIN

Proteins are the most complex substances that we study in the eighth grade. While we have learned many characteristics of carbohydrates and fats through our explorations, we can only gain a rudimentary knowledge of proteins by the lab procedures we perform.

Plants are able to synthesize starch, cellulose, and fat from the sugar created by photosynthesis, but they also need nitrogen to be able to create proteins. Even though the atmosphere is a vast reservoir of nitrogen, it is not in a form that plants can use. Instead, plants must absorb nitrogen that has been transformed into nitrates and nitrites by certain types of soil bacteria and archaea.

The plant uses the proteins it creates for various cell structures and functions, but most of the protein is concentrated in its seeds. The endosperm of the seed protects the embryo and contains the food reserves that the young plant will need after it germinates. This is why grains and legumes are so nutritious.

Human beings meet their protein needs by eating plants and animals. Animal protein is "complete," which means that it provides all the amino acids a person needs, but many plant proteins are "incomplete." This means that if a person chooses to be a vegetarian, she needs to make sure that the grains and vegetables she eats provide all of the essential amino acids for optimum health.

When discussing vegetarianism, it should be pointed out that the transformation of plant matter to animal protein is relatively inefficient: Chickens and pigs convert feed into meat at the rate of 2:1 or 3:1 (2–3 pounds of grain consumed to 1 pound of meat produced), sheep at about 6:1, and beef cattle as high as 20:1.

Of all the food substances, proteins are the quickest to spoil. Historically, people preserved proteins by smoking, salting, fermenting, and in cold climates, by chilling. More recently, refrigeration and the use of artificial preservatives have been used to keep proteins from spoiling.

Because proteins have a complex chemical makeup, we will primarily work with albumen (egg white) and gelatin (made from an animal protein called collagen), which are relatively simple proteins.

Gelatin is widely used by the food industry as a thickener, stabilizer, and texturizer. It is also used by pharmaceutical companies to make capsules easier to swallow, and as a conditioner and moisturizer in cosmetics. Collagen makes up about a third of the proteins in our body, providing elasticity and strength to our connective tissues. As we age, collagen production slows down, causing skin to sag and joints to become less supple.

In their lab work, students will begin by examining different proteins' relationship to water. They will find that albumen and gelatin behave slightly differently; albumen is partly soluble in cold water, while gelatin is fully soluble in cold water. Albumen coagulates in hot water, while gelatin forms a gel.

In terms of their relationship to heat, both albumen and gelatin undergo incomplete combustion and produce pungent odors. Students will notice that the proteins coagulate before burning; this corresponds to meat becoming firmer when cooked.

By mixing albumen or milk with hydrochloric acid, students gain a sense of what happens in their stomachs: the protein curdles and becomes partly solid. When protein is mixed with sodium hydroxide, they will see that the alkali breaks down the protein in a process similar to the one in our intestines.

As a special project, a student or lab group can learn how to test for protein and then share their findings with the class. The chemical test for protein is called the biuret test and is similar to the Fehling's and Benedict's tests for sugar, only it does not need to be heated, and the protein turns the solution lavender rather than copper-colored.

As a culmination to the chemistry block, students can have one final lab where they perform a series of tests to see if they can accurately identify a dozen or so unidentified substances. This kind of practical lab exam is a wonderful way to end the block, and students will be proud to demonstrate how much they have learned.

TOPICS

- The nature and use of proteins
- Protein interactions with water
- Protein interactions with heat
- Protein interactions with acids and bases
- Cooking proteins

CHARACTERISTICS

- Proteins come from animal and vegetable sources.
- When albumen is whipped, it traps air and becomes stiff.
- Albumen is slightly soluble in cold water and coagulates in hot water.
- Gelatin is soluble in cold water and creates a gel in hot water.
- When proteins are heated, they eventually combust, emitting a strong odor and producing a carbon ash.
- Acids cause proteins to coagulate; bases cause them to liquefy.

DAY	DEMONSTRATIONS AND ACTIVITIES	CHARACTERISTICS
1	■ The History & Use of Proteins ■ Examining Proteins ■ Albumen & Air	■ Proteins come from animal and vegetable sources. ■ When albumen is whipped, it traps air and becomes stiff.
2	■ Proteins & Water	■ Albumen is slightly soluble in cold water and coagulates in hot water. ■ Gelatin is soluble in cold water and creates a gel in hot water.
3	■ Heating Proteins ■ Making Meringues or Egg Drop Soup	■ When proteins are heated, they eventually combust, emitting a strong odor and producing a carbon ash.
4	■ Interactions of Proteins with Acids & Bases ■ Making Ricotta Cheese ■ Testing for Protein	■ Acids cause proteins to coagulate; bases cause them to liquefy.
5	■ Testing Unknown Substances to Determine their Identity	

A note on scheduling: If you have budgeted four weeks for the block, you will be able to spend the final week on proteins. If you only scheduled three weeks, I recommend that you spend two days on fats and two days on proteins. Choose the demonstrations and activities that seem most important and meaningful for your students. Reserve the last day of the block for the chemistry lab final.

DAY 1

PRESENTATION

1. The History & Use of Proteins

If you haven't already presented the development of culture from the hunter-gatherer stage to the establishment of settlements and cities, this would be a good opportunity to do so. Throughout human evolution, the ways that people have satisfied their need for proteins have determined a great deal about how society is organized.

Sources of vegetable and animal protein should be described, and the various ways of obtaining those proteins should be summarized.

If the topic of protein in diet and health was covered in the seventh-grade study of nutrition, it can be reviewed; if not, it can be considered now.

RELATED TOPICS

- Differences between plant and animal proteins
- Arguments for vegetarianism: ethical, ecological, and health
- The nitrogen cycle

ACTIVITY

2. Examining Proteins

MATERIALS

examples of animal proteins, such as beef, pork, chicken, fish, shellfish, milk, and eggs

examples of plant proteins, such as grains, legumes, and nuts

Characteristic: Proteins come from animal and vegetable sources.

PROCEDURE

Have the students recommend how best to group and categorize the proteins for purposes of comparison and contrast. Students should examine the proteins in terms of their physical qualities and should record their observations.

RELATED TOPICS

- Genetic engineering
- Factory farming

DAY 2

DEMONSTRATION

3. Albumen & Air

Characteristic: When albumen is whipped, it traps air and becomes stiff.

Beating egg whites reveals the tendency of proteins to become firmer.

PROCEDURE

- Separate an egg and put the white (albumen) in a medium bowl.
- Separate another egg and put the white in a beaker.
- Whip the egg white in the bowl until it form stiff peaks.
- Compare the whipped egg white with the liquid egg white.

SAFETY

The egg whites can be safely disposed of in the trash or compost.

RELATED TOPIC

Amino acids

EQUIPMENT

medium beaker

medium bowl

eggbeater, whisk, or electric mixer

MATERIALS

eggs

LAB ACTIVITY

4. Proteins & Water

Characteristic: Albumen is slightly soluble in cold water and coagulates in hot water.

EQUIPMENT

4 medium beakers
2 small beakers
stirring spoon or rod
electric kettle

MATERIALS

eggs
water
powdered gelatin

PROCEDURE

- **Setup**
  - Fill two beakers half full with cold tap water or refrigerated water and two beakers half full with water heated to boiling in an electric kettle.
  - Separate an egg and divide the albumen between two small beakers.
- **Albumen in cold water**
  - Put half the albumen in the beaker with cold water and observe its interaction with the water. Does it dissolve, float, sink, etc.?
  - Stir the albumen to see if it mixes with the water.
- **Albumen in hot water**
  - Put the remaining albumen in the beaker with hot water and observe its interaction with the hot water.
  - Stir the albumen.
- **Gelatin in cold water**
  - Pour a half packet of gelatin into the beaker with cold water.
  - Stir, and observe its interaction with the water.
- **Gelatin in hot water**
  - Pour the other half of the packet of gelatin into the beaker with hot water.
  - Stir and observe its interaction with the water.

SAFETY

- Students can work alone or in groups of two or three.
- The egg shells, albumen, and gelatin can be safely disposed of in the trash or compost.

RELATED TOPICS

- Uses and role of egg whites in cooking
- How gelatin is made
- Uses of gelatin in cooking
- Plant-derived gelatin
- Cosmetic and pharmaceutical uses of gelatin

DAY 3

LAB ACTIVITY

5. Heating Proteins

Characteristic: When proteins are heated, they eventually combust, emitting a strong odor and producing a carbon ash.

As students discovered when they burned hair, wool, leather, or a feather in seventh grade, proteins burn with a characteristic odor. Like carbohydrates and fats, proteins contain carbon, hydrogen, and oxygen. In addition, they contain nitrogen, and sometimes other elements, which are revealed in the odors created by combustion.

PROCEDURE

- **Heating albumen**
  - Set up the sand tray and chafing dish fuel.
  - Separate an egg and put some albumen in a small beaker for each lab group.
  - Heat a small amount of albumen (about ½ teaspoon) in the deflagration spoon.
  - Observe the coagulation, color changes, combustion, and the ash after it has burned.
- **Heating gelatin**
  - Using a new spoon, heat a tiny amount (¼ or ⅛ teaspoon) of gelatin.
  - Observe the color changes, combustion, and ash.

EQUIPMENT

sand tray
small beaker
2 deflagration spoons
pliers or tongs

MATERIALS

chafing dish fuel
matches
eggs
gelatin

SAFETY

- Students should work in groups of two or three.
- A window fan should be used to vent the odors.
- Very small amounts of protein should be heated because of the odors.
- The burned albumen and gelatin can safely be disposed of in the trash.
- The chafing dish fuel can be extinguished by using a pliers or tongs to replace the lid.

RELATED TOPICS

- Dietary and nutritional aspects of egg yolks and egg whites
- Dietary and nutritional aspects of collagen

COOKING ACTIVITIES

6. Making Meringues or Egg Drop Soup

Both of these activities will give students a good sense of how egg proteins react to heating

EQUIPMENT & MATERIALS FOR MERINGUE

electric mixer
medium bowl

---

4 large egg whites, at room temperature
¼ teaspoon cream of tartar
1 cup sugar
¾ teaspoon vanilla extract

PROCEDURE

(Recipes adapted from the website of America's Test Kitchen.)

- **Meringue Cookies**
  - Adjust the oven racks to upper- and lower-middle positions and heat the oven to 200 degrees.
  - Line 2 large baking sheets with parchment paper.
  - Beat the egg whites with an electric mixer at medium-low speed until they are opaque and frothy, about 30 seconds.
  - Add cream of tartar, increase the mixer speed to medium-high, and beat the egg whites until white and thick, about 90 seconds.
  - Slowly sprinkle in ½ cup of sugar and continue to beat until it is incorporated, about 60 seconds. Add vanilla.
  - Reduce speed to lowest possible setting, sprinkle in remaining

½ cup sugar, and mix until the sugar is incorporated.
  - Place dollops of meringue, about the size of a walnut, 1 inch apart, fitting 3 rows of 5 cookies on each sheet.
  - Bake for 1½ hours or until cookies have smooth, dry, and firm exteriors.
  - Turn oven off and allow cookies to cool in oven for several hours.
- **Egg Drop Soup**
  - Bring the stock to a simmer in a large saucepan over medium-high heat.
  - Add the soy sauce and salt to taste.
  - Combine the cornstarch and 2 tablespoons water in a small bowl and stir until smooth.
  - Whisk the cornstarch mixture into the stock until it thickens slightly, about 1 minute.
  - Stir in the scallions and cilantro.
  - Stir the stock so that it is moving in a circle, and while stirring, pour the eggs into the stock in a slow, steady stream so that ribbons of coagulated egg form, about 1 minute.
  - Let the eggs stand in the stock without mixing until they are set, less than 1 minute.
  - Once they have set, break up the egg ribbons with a fork.
  - Serve immediately.

SAFETY

If at all possible, cooking activities should not be done in a classroom that is used for chemistry. If there is no other alternative, thoroughly clean all surfaces, equipment, and utensils before beginning the activity.

EQUIPMENT & MATERIALS FOR SOUP

cutting board
knife
medium saucepan
whisk
fork

---

2 quarts chicken stock
1 tbsp. soy sauce
salt
2 tbsp. cornstarch
2 medium scallions, chopped fine
2 tbsp. fresh cilantro leaves, minced
4 large eggs, beaten

DAY 4

DEMONSTRATION

7. Interactions of Proteins with Acids & Bases

Characteristic: Acids cause proteins to coagulate; bases cause them to liquefy.

Prior to this demonstration, students should be reminded of what they learned about the effects of stomach acids and intestinal juices on proteins: acids "cook" food substances, while bases "digest" them.

EQUIPMENT

small bowl
2 plastic pipettes
watch glasses

MATERIALS

concentrated hydrochloric acid
milk
egg
sodium hydroxide
water

PROCEDURE

- **Setup**
  - Carefully pour a little concentrated hydrochloric acid into a small beaker and cover it with a watch glass.
  - Mix a concentrated solution of sodium hydroxide in a small beaker and cover it with a watch glass.
- **Curdling milk**
  - Pour ¼ cup of milk in a bowl.
  - Using a pipette, add three drops of hydrochloric acid into the milk and stir.
  - Observe the reaction, if any.
  - Continue to drop more acid into the milk until it curdles completely.
- **The effect of acid on albumen**
  - Put a small amount of albumen on the watch glass.
  - Using a pipette, place three drops of hydrochloric acid on the albumen and observe.
  - Continue to drop more hydrochloric acid on the albumen until it is fully coagulated.
- **The effect of base on albumen**
  - Using a pipette, place three drops of sodium hydroxide solution

on the "cooked" albumen.

- Continue to place more drops of sodium hydroxide until the albumen returns to its fluid state.

SAFETY

- Everyone needs to wear goggles, gloves, and aprons when working with hydrochloric acid and sodium hydroxide.
- When preparing the sodium hydroxide solution, be sure to put the sodium hydroxide into the water. Do not pour the water into the sodium hydroxide.
- Keep the beakers with hydrochloric acid and sodium hydroxide covered with watch glasses and in a place where they will not get jostled.
- The milk that has been curdled with hydrochloric acid can be neutralized with baking soda, diluted, and safely flushed down the sink.
- The albumen that has been mixed with hydrochloric acid and sodium hydroxide can be diluted and safely flushed down the sink.
- Any remaining acid or base can be neutralized and safely flushed down the sink.
- The pipettes can safely be disposed of in the trash.

RELATED TOPIC

The structure of proteins

COOKING ACTIVITY

8. Making Ricotta Cheese

The curdling of milk protein is an essential step in making cheeses. Rennet—the lining of a calf's stomach—is often used to curdle the milk, but acids can also be used.

EQUIPMENT

Dutch oven or large heavy saucepan

instant-read or candy thermometer

rubber spatula

colander

MATERIALS

1 gallon pasteurized whole milk

⅓ cup fresh lemon juice

¼ cup distilled white vinegar, plus extra as needed

2 teaspoons salt

cheesecloth or muslin

PROCEDURE

(Recipe adapted from the website of America's Test Kitchen.)

- *Note: For best results, don't stir the milk too hard, and be very gentle with the curds once they form.*
- Line the colander with a triple layer of cheesecloth and place it in a sink.
- Combine lemon juice and vinegar in a liquid measuring cup and set aside.
- Add salt to the milk and heat in a Dutch oven over medium-high heat. Stir frequently with the rubber spatula to prevent scorching, until the milk registers 185°F.
- Remove the Dutch oven from the heat and slowly stir in the lemon juice mixture until it is fully incorporated and the mixture curdles, about 15 seconds.
- Let sit undisturbed until the mixture fully separates into solid curds and translucent whey, 5 to 10 minutes.
- If curds do not fully separate and there is still milky whey in the Dutch oven, stir in extra vinegar, 1 tablespoon at a time, and let sit another 2 to 3 minutes, until curds separate.
- Gently pour mixture into the prepared colander.
- Let sit, undisturbed, until the whey has drained from the edges of cheese but the center is still very moist, about 8 minutes.
- Gently transfer the cheese to large bowl, retaining as much whey in the center of cheese as possible.
- Stir well to break up large curds and to incorporate the whey.
- Refrigerate ricotta until cold, about 2 hours.
- Stir cheese before using.
- Ricotta can be refrigerated for up to 5 days.

SAFETY

If at all possible, cooking activities should not be done in a classroom that is used for chemistry. If there is no other alternative, thoroughly clean all surfaces, equipment, and utensils before beginning the activity.

RELATED TOPICS

- Types of cheese and how they are made
- The fermenting process and fermented milk products

LAB ACTIVITY

9. Testing for Protein

Characteristic: Biuret solution reveals the presence of proteins.

This activity can be done as an independent project by students who want to explore another chemical test.

It can be done with a series of labeled and unlabeled substances to make it more interesting and challenging.

PROCEDURE

- Pour about 1 inch of biuret solution into a test tube.
- Put a small amount of protein in the solution.
- Shake the test tube gently and allow to sit for 5 minutes.
- Observe whether the solution turns purple.

SAFETY

- Students should wear goggles and gloves for this activity.
- Check local ordinances regarding safe disposal.

EQUIPMENT

test tubes
test tube stands
small beakers or shot glasses
paper baking cups

MATERIALS

various proteins, such as:
albumen
gelatin
powdered collagen
powdered milk

DAY 5

As a final experience, students should have the opportunity to put all the knowledge and skill they have gained to use through a chemistry final lab exam. I have assigned this kind of final exam to many classes, and it has always been successful. Although it requires a fair amount of preparation, it is well worth the time and effort.

The tests take up to an hour, so plan to use the entire main lesson time and leave time at the end for cleanup and for comparing results.

You should decide whether to have students work individually or in groups. Groups should have no more than three members.

Students will need to take notes to keep track of their observations and conclusions. (If students work in groups, a note taker can be chosen by each group.)

Students should keep their conclusions secret during the exam so that other groups don't pirate their results.

Students should wear goggles and gloves for the entire exam and show that they can perform all the procedures safely.

During the exam, you should go around the room and note the level of each student's laboratory skill and confidence.

CHEMISTRY FINAL LAB EXAM

10. Testing Unknown Substances to Determine Their Identity

PROCEDURE

- **Setup**
  - Prepare 10 to 12 substances for students to test.
  - Whenever possible, use powders that look similar to each other, so the students cannot recognize the substances based on their prior work.
  - Put each substance in a small paper baking cup, using a Sharpie to label the paper with a number on its side. Keep a list that identifies the substances by their numbers.
  - Put all the baking cups on a tray for each lab group.

- **Process**
  - Review with the students the tests that are available:
    - observing the substance's physical characteristics
    - heating and combusting
    - mixing with cold and hot water
    - Fehling's solution test for simple sugars
    - iodine test for starch
  - You can set an extra challenge by limiting how many times they can use a specific test. For instance, allow only four substances to be tested with Fehling's solution.
  - Students may realize that they will probably be able to recognize a substance by heating it. Encourage them to verify their conclusion by another test.
  - When the exam is over and everything is cleaned up, review the results together. If a group has made an error, see if they can figure out what may have led them to their result.

SAFETY

- The chafing dish fuel can be extinguished by using a pliers or tongs to replace the lid.

EQUIPMENT

Each group needs the following:

1 sand tray

1 can of chafing dish fuel

pliers or tongs

several deflagration spoons

matches

splints

laboratory scoop

2 watch glasses

1 test tube rack

8–10 test tubes

1 test tube with about 4″ of Fehling's A solution

1 test tube with about 4″ of Fehling's B solution

tincture of iodine

MATERIALS

Suggested substances:

sugars: fructose, glucose, and sucrose

starches: cornstarch and arrowroot

fats: stearic acid and finely grated paraffin

proteins: collagen and powdered egg whites

non-foods: baking soda, calcium hydroxide, salt, or other white compounds

Appendix 1
CHEMICALS & EQUIPMENT

CHEMICALS

- barium chloride (seventh grade)
- Benedict's solution (eighth grade)
- biuret solution (eighth grade)
- calcium chloride (seventh grade)
- calcium hydroxide (seventh grade)
- calcium oxide (seventh grade)
- citric acid (seventh grade)
- copper chloride (seventh grade)
- copper strips (seventh grade)
- Fehling's solutions A and B (eighth grade)
- hydrochloric acid, concentrated (muriatic acid can be substituted) (seventh and eighth grades)
- lithium chloride (seventh grade)
- lycopodium powder (seventh grade)
- magnesium ribbon (seventh grade)
- potassium chlorate (seventh and eighth grade)
- potassium chloride (seventh grade)
- potassium dichromate (seventh grade)
- sodium chloride (can substitute with Kosher salt) (seventh grade)
- sodium hydroxide pellets (seventh grade)
- sodium hydroxide powder (household lye can be substituted) (seventh and eighth grades)
- strontium chloride (seventh grade)
- sulfur, powdered (seventh grade)
- sulfuric acid, concentrated (seventh and eighth grades)
- zinc strips (seventh grade)

LABORATORY EQUIPMENT & GLASSWARE:

- beakers, small, medium, large
- Bunsen burner
- butane lighters
- chafing dish fuel
- deflagration spoons
- Erlenmeyer flasks, 125 ml, 250 ml, 500 ml, 1,000 ml
- filter paper
- funnel
- hot plate
- laboratory stands
- ladle, small
- MAPP gas/oxygen torch with regulator
- pipettes
- plastic tubing
- ring clamps
- safety screen
- scoops
- stoppers, 1-holed and solid
- test tube clamps
- test tube racks
- test tubes
- tongs
- utility clamps
- watch glasses
- wire screens

Appendix 2

COMMON LABORATORY PROCEDURES

1. POURING LIQUID FROM A FLASK INTO A BEAKER

This procedure can also be used to transfer liquids from one beaker to another.

- Hold the beaker firmly on a stable surface with one hand.
- Hold the flask firmly with the other hand.
- Rest the lip of the flask on the rim of the beaker.
- Tilt the flask to pour the liquid.
- Do not release the beaker until all the liquid has been poured.

2. TRANSFERRING LIQUIDS FROM A FLASK, BEAKER, OR TEST TUBE INTO A TEST TUBE

- Hold the test tube with one hand so that its base is on a stable surface.
- Alternatively, place the test tube in a test tube holder.
- Hold the container holding the liquid with the other hand.
- Rest the lip of the container on the rim of the test tube.
- Tilt the container to pour the liquid.
- Do not release the test tube until all the liquid has been poured.

3. FILTERING A SUBSTANCE FROM A LIQUID SOLUTION OR MIXTURE

- Place a funnel on a flask or beaker.
- Fold the filter paper in half, and then in half again.
- Place the filter paper in a funnel.
- Insert a stirring rod in between the filter paper and the funnel to aid filtration.
- Slowly pour the solution or mixture into the filter paper.

4. DILUTING ACIDS OR BASES

- Fill a beaker a quarter to half full of distilled water.
- Follow the procedures described in #1 with the liquid acid or base.
- When mixing a solid acid or base with water, carefully pour the acid or base into the water.

5. FILLING A PIPETTE WITH LIQUID

- Have the liquid ready in a small beaker.
- Squeeze the pipette firmly; air should be expelled.
- Continue to hold the pipette firmly squeezed and insert it into the liquid so that the tip is just above the bottom of the beaker.
- Release the pressure and allow the pipette to fill itself.
- Do not squeeze the air out of the pipette while the tip of the pipette is in the liquid.

6. POURING ACID OR BASE ON A SUBSTANCE

- Pour acid or base into a small beaker using procedures described in #1.
- Cover the beaker with a watch glass until ready to use.
- Place the substance in a beaker or on a watch glass.
- Use a pipette to draw up acid or base following the procedures in #5.
- Hold the pipette about ½ inch above the substance.
- Carefully squeeze the pipette to drop acid or base on the substance.

7. HEATING SOLID OR GRANULAR SUBSTANCES IN A TEST TUBE ATTACHED TO A LABORATORY STAND

- Attach the test tube to a laboratory stand with a test tube clamp or utility clamp.
- Fill the test tube with a small amount of the substance. (It should be no more than a quarter full).
- Adjust the test tube so that it slants at about a 45-degree angle.
- Adjust the height of the test tube, depending on how intense you want the flame to be.
- **Please note:** *The demonstrations of heating a gummy bear with potassium chlorate and of melting and burning paraffin require the test tube to be held upright when heated.*

8. USING CHAFING DISH FUEL

- Place the can of chafing dish fuel on a stable, fireproof surface.
- Pry the lid off the can with a knife, spoon, or screwdriver.
- Use a butane lighter or a match to light the fuel.
- To extinguish the flame, hold the top of the can with pliers or tongs and place the lid on top of the opening.
- Allow the can to cool before moving it.
- The can of chafing dish fuel does not burn as well when it is low on fuel.
- Fuel can be transferred from other cans that are low. Be careful that all the fuel ends up inside the can. If it spills, it is a fire hazard.
- Empty cans of chafing dish fuel can safely be disposed of in the trash.

9. HEATING LIQUIDS IN A TEST TUBE HELD BY A TEST TUBE CLAMP

- Place the chafing dish fuel on the sand tray.
- Fill the test tube no more than a third full of the liquid.
- Grip the test tube with the test tube clamp so that the clamp is a little below the rim of the test tube.
- Light the chafing dish fuel.
- Hold the test tube so that it slants at about a 45-degree angle, pointing away from everyone.
- Heat the top of the liquid first. As it starts to bubble, gently move the test tube so that more of the liquid heats up.

10. BURNING SUBSTANCES BY HOLDING THEM OVER A FLAME WITH A TEST TUBE CLAMP OR TONGS

- Place the alcohol lamp or chafing dish fuel in a sand tray.
- Light the alcohol lamp or chafing dish fuel.
- Grip the substance with a test tube clamp or tongs and hold it just above the flame until it ignites.
- Once it is lit, move the substance away from the flame to observe its combustion.
- When the combustion is completed, cap the alcohol lamp or chafing dish fuel.

11. BURNING SUBSTANCES IN A DEFLAGRATION SPOON

- Place the chafing dish fuel in a sand tray.
- Light the alcohol lamp or chafing dish fuel.
- Fill the spoon with a small amount of the substance.
- Hold the spoon just above the flame until it ignites.
- Once it is lit, move the spoon away from the flame to observe its combustion.
- When the combustion is completed, cap the chafing dish fuel using pliers or tongs.

12. USING A BUNSEN BURNER

- Make sure that the surface where the burner will be placed is clear.
- Inspect the rubber hose to make sure that it has not been damaged.
- Check that the gas at the supply main is turned off.
- Insert the ends of the hose into the supply main and into the burner.
- If using a propane tank, check the clamps attaching the hose to the tank and the hose to the burner to make sure that all the connections are tight.
- Adjust the collar so that it almost closes off the air ports, leaving just a little gap for air to enter.
- If the Bunsen burner has a needle valve, check the valve to see if it is completely closed. If not, turn it until it is closed.
- Open the gas main. You should hear no gas escaping. If you do hear it, shut the gas off and check the needle valve to make sure it is closed.
- Have the lab striker or lighter ready and open the needle valve a half turn. You should hear the gas escaping.
- Hold the lab striker or lighter 1–2 inches above the barrel and squeeze the striker to create a spark or pull the trigger on the lighter.
- As soon as the flame ignites, move the striker or lighter away.
- Use the needle valve to adjust the height of the flame.
- Use the collar to adjust the flame's temperature.
- Always move the burner by the base; the barrel will be hot.
- To extinguish the burner, adjust the collar to create a cooler flame.
- Turn off the gas supply at the main.
- Close the needle valve.

Appendix 3
CHEMICAL STORAGE

Chemicals should be segregated and stored in separate cabinets designed for specific categories. Because this manual requires so few chemicals, it may be preferable to use multi-use storage cabinets or closets.

1. CORROSIVE ACIDS

The acids used in this manual that need to be stored in a separate, corrosion-resistant storage cabinet include:

- hydrochloric acid
- sulfuric acid

The acids used in this manual that do not need a special storage cabinet include:

- vinegar
- citric acid

2. CORROSIVE BASES

The bases used in this manual that need to be stored in a separate corrosion-resistant storage cabinet include:

- sodium hydroxide
- calcium oxide
- calcium hydroxide

The bases used in this manual that do not need to be stored in a special storage cabinet include:

- baking soda
- Epsom salts

3. FLAMMABLES

The flammable substances that need to be stored in a flammables cabinet include:

- sulfur
- potassium chlorate
- magnesium ribbon
- flash paper
- flash cotton
- alcohol
- hand sanitizer
- chafing dish fuel
- propane fuel
- MAPP gas and oxygen cylinders and regulator

The flammable substances that do not need special storage include:

- steel wool
- lycopodium powder
- stick charcoal
- dried parts of plants
- cotton
- pieces of cloth
- sugars
- starches
- oils
- proteins

4. TOXIC SUBSTANCES

The toxic substances that need to be stored in a special cabinet include:

- bleach
- hydrogen peroxide
- potassium dichromate
- chlorides of salts
- Fehling's solutions A and B
- Benedict's solution
- biuret solution
- tincture of iodine

Appendix 4

EXCERPTS FROM THE WORK OF EUGEN KOLISKO

Eugen Kolisko, one of the teachers at the first Waldorf school, worked with Rudolf Steiner. Dr. Kolisko described his approach to chemistry in the seventh and eighth grades in *Natural History* and in his pamphlet *First Lessons in Chemistry*. I have provided excerpts from these two works in the hope that they will inspire the reader to read the originals in their entirety and to ponder the insights that Dr. Kolisko shared.

COMBUSTION

Excerpted from *Natural History*:

> Rudolf Steiner's constant aim was to open up new conceptions of the being of man and his relation with the universe. The teacher himself is educated by these things; his being is changed in a certain sense, and having imbibed such points of view in his preparation, all the lessons he gives will be quite different in character....
>
> In the course of my earlier work in laboratories I had wrestled hard with the problem of how to understand chemical phenomena in such a way that they reveal their nature simply of themselves, each throwing light on the other. I used to call this mode of observation "phenomenological chemistry." It was an imperfect attempt to understand nature in the sense of Goethe's words: "Let us look for nothing behind the phenomena, for they themselves are the teaching." Nevertheless, the methods I had adopted in this study were a help to me now when it was a question of working out all that the astounding ideas given by Rudolf Steiner brought to light. In connection with the plan of teaching, he had said laconically: "The best way to teach the elementary principles of chemistry is to begin with the phenomena of combustion." For years, I have been working to fathom the depths of a sentence like this. Be that as it may, I began by

trying to show what a flame really is: how on the one side light and warmth stream out into the world—in other words, how the imponderable elements of light and warmth return to the heavens; and on the other side, how the ashes fall away—the dead, hardened, wholly "earthly" ashes. Between these two processes there is the air, giving life to the flame; and there, too, is the smoke wherein the air and the water are wrestling with each other. The whole is the flame, where the great cosmic antitheses are released from one another: light and warmth at one pole, and at the other, matter—solidified, salt-like and earthy. Everything that is living is combustible. The withering of the leaves in autumn is like a great flame soaring to the heavens, leaving ashes behind it. Spirit is freeing itself from matter. This is how we regard the fire process. A saying of Goethe sums it up very beautifully: "Whatever is laid hold of by the original force living in the fire does not remain either a deformation or as a burden on the Earth. Nay, it becomes ethereal and invisible, soaring upwards to its primal origin." In this sense we can speak to the children about the various phenomena of combustion and burning. We let nature herself be an illustration of the fire-process, in the blossoming of plants and in the withering of leaves. Wherever there is burning, there too we find expression of the great cosmic antitheses, and children learn to think about every fire phenomenon in this way.

ABOUT CALCIUM AND SALT FORMATION

Excerpted from *First Lessons in Chemistry.*

Understanding opposites is very important in chemistry. We began by discussing combustion, and the opposite of combustion is salt formation. Calcium provides the best example for presenting salt-forming phenomena. First we show the many different ways in which calcium appears in nature.... Next we ask, how did all these formations come about? Ultimately, they were all deposited by water. It takes a long time, for example, for stalactites to develop through the evaporation of lime-rich water. We also explain how the delicate shells of dead animals are constantly raining down out of seawater to form calcareous mud on the ocean floor. That is how chalk develops – out of the countless shells of tiny living things. In some cases, entire mountains, such as the chalk cliffs along the North Sea and the Baltic, formed out of chalk deposits. Shell limestone develops through a similar process. All of this takes a very long time. Limestone mountains develop very slowly. Calcium drops to the sea floor and is compacted and solidified. This process is very different from combustion. In combustion, we must first set the various plant parts on fire, and then they disperse as they are transformed into fire and smoke.

Now we bring an assortment of shells and stones into class – as many different ones as possible. They have all either been deposited under water or excreted by living things. When the solid element separates from the fluid element, everything proceeds very slowly and quietly, as when salt precipitates out of seawater. In fire, the activity of heat, light, and air predominates, and little solid substance remains. The processes of crystallization and precipitation of lime and other salts are totally different. It is true that lime develops from living things, but the process is not at all like fire, which consumes everything living and returns it to heaven. Instead, everything precipitates to earth out of the watery element of life. Here gravity and heaviness predominate. When children observe these phenomena over a longer period of time, they often discover these great contrasts for themselves....

The behavior of the class is quite different during these descriptions of salt formation than during previous discussions of fire. Fire is stimulating.... When we consider lime, the mood is different and more thoughtful. We are encouraged to ponder how this infinite variety of rocks developed over long periods of times. Crystallization takes time, and the process must not be disturbed if perfect crystals are to form....

We can then ask, if fire works in human blood and in the movement of our limbs, where do we find salt processes in the human body? They are concentrated in the head, where there is also the most bone. If lime were not deposited up there in our heads, clear thinking and calm understanding would never be possible for us. There would also be no bony scaffolding to support us, and the whole body would collapse. This is how we relate lime formation processes to the human being. We have seen that combustion processes are related to the lower part of the human body and its limbs; salt processes to the upper part, to the head. On the other hand, we have also laid the foundation for a perception of the fact that death constantly emerges from life, and in fact that is how most of the globe was formed. Anyone who has internalized this fact will never attempt to explain life on the basis of dead matter. Our observations about lime, therefore, are connected to the human being on the one hand and to the world on the other.

In the next lesson, we can take our observations of lime in a different direction. We "calcine" a piece of limestone by heating it, preferably under forced air inflow such as with bellows. We allow the burnt lime to cool and then pour water on it. It fizzes and heats up. We repeat the experiment with a larger quantity of burnt lime. It swallows the water greedily. Relatively large amounts of water disappear into the burnt lime without a trace. After a while, it begins to smoke and generates a lot of heat. Burnt lime appears to be deficient in water. Through the activity of fire, it has been completely separated from the water from which it originated, which is why it then absorbs water so greedily. The thoroughly "animal" nature of lime is evident even long after its last cycle through animal bodies. Adding even more water produces a milky liquid known as milk of lime or whitewash. If the mixture is left undisturbed, the slaked lime settles, but some of it remains in solution in the clear liquid (limewater). This limewater turns litmus paper blue, whereas the original, unburnt lime did not. A liquid that turns litmus paper blue is called a base. All bases also share a very characteristic taste, just as acids do. Heating limestone and slaking it with water produces a base....The gas that escapes can also be mixed with water. Carbon dioxide gas gives water a sour taste, and litmus paper turns red in it. This is how carbonic acid develops from limestone.

We have now introduced the children to the concepts of acid and base. Limestone is actually a salt, the product of a salt-forming process. Limestone changes when exposed to fire, releasing carbon dioxide gas. Burnt lime remains behind, and when water is added to it, a base is formed. Adding water to the substances separated by fire results in an acid and a base. Their opposite character becomes evident when each is combined with water....

It is very important to avoid the usual approach of beginning with the acid and the base and deriving the salt from them. Instead, we must take the opposite approach, which is the natural way, because lime already exists everywhere in nature, whereas the acid and the base are only produced later and by artificial means. The opposites inherent in lime become apparent only when we intervene to chemically alter the salt.... In chemistry, we begin with the totality and allow the parts to emerge from it. This is an important point. Even at this level of instruction, teachers must sense that a "chemical compound" cannot be seen as simply the sum of its parts or the chemical sum of its elements. It is something new. In many cases, it is actually the natural, original material.

Thanks to the Kolisko Archive Publications for their kind permission to quote these passages.

Bibliography

Bojarsky, Mikko & Antje Borjarsky. *A Demonstration Manual for Use in the Waldorf School 7th Grade Chemistry Main Lesson*. Mikko Bojarsky, 2005.

Bojarsky, Mikko. *A Demonstration Manual for Use in the Waldorf School 8th Grade Chemistry Main Lesson*. Mikko Bojarsky, 1999.

Fairman, Eric K. *A Path of Discovery: Grades 6, 7, 8*. Sidney, Australia: Eric Fairman, 2001, 2002.

Faraday, Michael. *The Chemical History of a Candle*. New York: Dover Publications, 2003.

Julius, Fritz. *The World of Matter and the Education of Man*. Forest Row, Sussex: Steiner Schools Fellowship Publications.

Kennish, Graham. *The Teaching of Chemistry in 7th & 8th Grade*. Gloucester, UK: 1997.

Kolisko, Eugen. *Elementary Chemistry*. Kolisko Archives, 1978.

Kolisko, Eugen. *Natural History*. Kolisko Archives, 1979.

Mitchell, David. *The Wonders of Waldorf Chemistry*. Chatham, NY: AWSNA, 2004.

Sonner, Robert, H. *Chemistry for Waldorf Middle Schools: Grades Six, Seven, & Eight – A Compendium of Phenomenological Experiments*. Mission Hills, CA: Robert H. Sonner, 2014.

Snijders, Leo. *Chemistry in Grade 7 of Waldorf Schools: A Main Lesson Block of 3 weeks: A Short Guide to Experiments on Fire, Lime, Acid, Base and Salt: based upon the work of Julius and von Mackensen*. Holland: Snijders, 1997.

Stockmeyer, E. A. Karl, *Rudolf Steiner's Curriculum for Waldorf Schools*. UK: Steiner Waldorf Schools Fellowship, 2001.

Wilkinson, Roy, *Teaching Physics and Chemistry*. Fair Oaks, CA: Rudolf Steiner College Press, 1978.

CHEMISTRY TEACHING REFERENCES IN RUDOLF STEINER'S WRITINGS:

A Modern Art of Education. lecture 10. Great Barrington, MA: Anthroposophic Press, 2004.

Practical Advice to Teachers. lecture 8. Great Barrington, MA: Anthroposophic Press, 2000.

Soul Economy and Waldorf Education. lecture 11. Great Barrington, MA: Anthroposophic Press, 2003.

The Kingdom of Childhood. lectures 3 and 7. Hudson, NY: Anthroposophic Press, 1995.

The Renewal of Education. lectures 8 and 10. Great Barrington, MA: Anthroposophic Press, 2001.

The Spiritual Ground of Education. lecture 6. Great Barrington, MA: Anthroposophic Press, 2004.

Education for Adolescents. lectures 1 – 4. Hudson, NY: Anthroposophic Press, 1996.

Made in United States
Orlando, FL
05 November 2026

90231408R00118